ON MY MIND

Reflections on Animal Behavior and Learning

KAREN PRYOR

karen
pryor
CLICKER TRAINING

On My Mind
Reflections on Animal Behavior and Learning

Karen Pryor Clicker Training
Sunshine Books, Inc.
49 River Street, Suite 3
Waltham, MA 02453 USA
U.S. (Toll Free) 800-472-5425
781-398-0754

www.clickertraining.com

For information about special discounts for multiple-copy purchases, please contact
Karen Pryor Clicker Training sales:
U.S. (Toll Free) 800-472-5425 or 781-398-0754 or wholesale@clickertraining.com.

Editing: Nini Bloch
Cover/book design: Rosamond Grupp
Cover photo: Nils Schlebusch

First edition published 2014

Printed in the United States of America

ISBN-10: 1-890948-99-3
ISBN-13: 978-1-890948-99-3

Library of Congress Control Number: 2014914450

To Misha
1999–2014

Table of Contents

Preface

The essays in this book were written between 2002 and 2014. They were called "Letters from Karen" and sent out more or less monthly from www.clickertraining.com, the website of the publishing and teaching company I founded, now called Karen Pryor Clicker Training (KPCT). At first these letters were about company-sponsored events and products. As our audience grew and more trainers' questions came in, the letters evolved into training discussions. Then they also became a place for me to share my own thoughts or opinions—what people now would call a blog. I produced more than a hundred of these letters over the years.

Writing these pieces was a curiously solitary experience. Letters went out every month to somewhere between fifteen and thirty thousand people, but I almost never heard another word about them. Did people actually read them? Hello? This year, 2014, one person came up to me at a conference and told me she loved the coffeepot story. A reader! Hurray! Meanwhile I remained my own fan. I had fun writing "Too many Butterflies" and "Bad Bob," and others, and I enjoyed rereading the letters for this book.

However, until this collection was assembled, I didn't recognize what a great podium this monthly duty became. From here I could dissect perpetually confusing questions such as the use of extra treats ("Jackpots: Hitting it Big") or the frequency of unintentional punishers ("A Scaredy-Cat Dog;" "Why I Hate the Long Down"). I could demonstrate just how to interact with wild animals in the wild using our positive principles ("The Dolphin Witch;" "Clicking Raptors in the Saguaro""). I could explore new areas of marker-based training, unfamiliar to traditional trainers and not seen in the scientific literature either, such as the existence of highly developed creativity in some animals not generally regarded as intelligent ("101 Things to Do with a Polar Bear;" "The Panda Game," which features a horse named Panda). In short, I could go off the main track and share some of the most exciting aspects of this technology as it zooms past the traditional worlds of laboratory, classroom, stables, and kennel. Each letter may not have produced much immediate feedback for me when it was written. But now I can see, as the training population has shifted gears over the last few years, that I may have planted a lot of seeds. And they grew.

My current work has taken me back to my science side: doing research, working with students, and publishing papers in the peer-reviewed literature. As this book goes to press, the learners I'm presently studying are not dogs or horses or dolphins, but medical students acquiring the skills needed to handle surgical tools. Yes, they learn fast, and yes, they use the clicker sometimes, and yes, it's tremendous fun for everyone. SEEKING systems are fully engaged and reinforcements, from praise to pizza, abound. Click!

Acknowledgments

I am grateful to Julie Gordon, KPCT web content manager and retail sales manager, who conceived of this project, selected the letters, negotiated the use of the great cover shot from photographer Nils Schlebusch, oversaw the design, and arranged for publication. I am equally grateful to editor Nini Bloch, who gave me insightful and very welcome guidance. Nini also organized the essays into something that made sense, crafted introductions and comments where needed, created the Find Out More! section, and brought the book to fruition.

The SEEKING System

K aren loves to play. Play takes curiosity, keen observation, understanding your partner, good timing, and a sense of humor. Karen has all those qualities. Maybe that's why she's such a superb clicker trainer.

She's also an explorer. Turns out that play is the purest form of exploration. And, recent scientific studies have shown that all animals are hard-wired to explore. It's not just the goodies that the curious get from exploring (most primitively, food, shelter, and sex) that drives them, it's the joy in the process that keeps us all going. We are all seekers, and that, more than anything else, is what clicker training taps in to and what makes it so much fun.

Pip greets his new friend Karen Pryor in proper goat fashion—by touching noses. Clicker training builds trust, enhances communication, and encourages play across species (see page 15). Photo: Steve Golson

1

Training on TV

In recounting a classic demo of clicker training that
Karen gave on a TV show using a shelter dog, she marvels at
the speed of the dog's epiphany. Such results astonish even a
Master Shaper who's had years of experience to
hone her focus, timing, and eye.

The rapidity with which dogs can convert from chronic, reactive behavior to thoughtful, attentive, operant behavior continues to amaze me. It's as if you give them a new tool kit and they just chuck the old methods out the window. There are two keys to making this conversion happen: having many small pieces of an appetizing primary reinforcer, and getting clickable moments going at a very high rate, many clicks a minute.

The first time I really pulled off that high rate of reinforcement was on television. I was in Chicago at a scientific conference. Steve Dale, a well-known radio and TV personality, invited me to do some clicker training during his weekly section on pets on the noon news. What could I do in five or ten minutes that would show up clearly on TV? I asked him to arrange to have one of the shelters bring in a large and friendly but out-of-control dog, a dog less than two years old. (Shelters see a lot of these. When the dog was a little puppy, it was so adorable, bouncing into everyone's lap and licking them madly on the face, that such behavior was not only permitted, but thoroughly reinforced. It's not so cute when the dog weighs 80 pounds and knocks down elderly relatives. Indeed, Steve had no problem locating a candidate for the show.)

I came to the TV studio with a clicker and a half-pound of cut-up roast chicken from a nearby deli. The dog arrived—a big, yellow, tail-wagging, leaping, face-licking, semi-Labrador—perfect! The cameras were running. The dog dragged the shelter volunteer onto the set and headed straight for me and my pieces of chicken. Click, treat. Ingesting roast chicken made him stand still for a moment; it may have been the first time in his life he tasted something that delicious. It may have been the first time that day he had actually stood still, too. I clicked and treated again.

I sent the shelter volunteer away, gave Steve the leash, and told him, "Follow the dog and keep the leash somewhat slack. You are just the emergency brake in case the dog decides to leave the building." I started walking slowly in a circle on the TV stage. When the dog jumped on me, nothing happened. When he made excursions away from me, nothing happened. When he jumped on Steve, nothing happened. Whenever he came near my left side, however, he got a click and a pea-sized piece of chicken. I kept walking, and the dog started staying with me more and more. I clicked and treated every two steps, and then every three. Soon the dog was walking quietly beside me for several steps. He began staring at my face; I timed my clicks to catch that attentiveness, too. Then I raised the bar a little, clicking every eight steps, and then every ten. Still the dog kept pace with me, followed by Steve Dale who was grinning and keeping the leash slack as instructed.

What would happen if I stopped walking now? Would the dog leave me? I took that chance, and stopped. The dog not only stopped, it actually sat. Click! Big treat. Someone in his past had taught him that behavior, and now it not only surfaced, it paid off for him. See, dog? You can make people do what you want, but in a new way, not the way you've been trying all this time. I moved next to Steve. How about sitting for Steve? Steve spoke to the dog. Instead of jumping on him, the dog sat in front of him. Yippee! I clicked, and Steve gave the treat.

My co-star and my friend

In five minutes, this now-model citizen and I walked off the set together, side by side, as if we'd been friends for years. The dog would need experience in other locations and with other people before being able to control himself at all times when walking on a leash. But already he had come a long way. He was learning to learn. He was calmer. He had a new interest in people. He was now looking at faces instead of only at hands and food. He was already more likely to get adopted, and the adoption was more likely to stick.

Giving a restless, excited shelter dog some active, operant behavior to do for reinforcement is a wonderful alternative to trying to suppress over-the-top behavior with correction and physical restraint. Go down to your local shelter and try it yourself!

The Dolphin Witch

*As a behavioral biologist, Karen is a keen observer of
all beings, from hermit crabs to grandchildren. In this story,
Karen's intimate knowledge of what makes dolphins tick
enables her to bewitch not only her subjects but,
apparently, her boat crew as well.*

My second husband, Jon Lindbergh, was an aquaculture expert with many clients in South America. I often traveled with Jon when he visited these clients. On this trip we were in Chile, visiting a salmon farm in a remote part of the southern archipelago. The area is where the Andes dip down into the sea, leaving a scattering of little islands that are actually flooded mountaintops.

Jon and I and several fish-farm workers were crossing a small bay in a workboat, heading for the salmon pens, when I spotted a few little dolphins in the distance. I knew what they were: *Cephalorhyncus eutropia* (the Chilean dolphin or, locally, the *tonino,* a shy and little-known marine mammal that lives in small groups along these sheltered shores).

I had been bickering with marine scientist Ken Norris (my scientific mentor) about the color of this species. He had seen them from shipboard and referred to them as the Chilean Black Dolphin, but I'd seen them around the fish farms, and they're definitely not black.

I wanted a closer look, but the guys said the animals would leave if we went toward them. I suggested calling them over instead. Dubious smiles all around, but they stopped the boat. I noticed some short lengths of aluminum pipe lying on the floorboards. I asked one of the men to hold one end of a piece of pipe in the water and bang on the other end with another piece; that would make a distinct sound underwater. I could have done the pipe work, but I needed my hands free for the camera.

The fish farmer started tapping out a nice little samba rhythm. Immediately, the dolphins stirred, turned, and came over to us. They were much smaller than good old Flipper, the Atlantic bottlenose—think cocker spaniel compared to golden retriever. They swirled around and under the boat, inspecting and almost touching the pipe.

These dolphins don't have beaks or snouts; their faces just come to a point like a toy animal. Cute! They proved to be light beige on top, with a cream-colored hourglass pattern on the belly (they obligingly swam upside down quite a lot). Most of them were also decorated with, of all things, a little white ring around the neck like a collar. I got some clear pictures.

Before the little dolphins lost interest, I sat down and told the guys that was enough. The samba rhythm stopped, and the dolphins left instantly.

I noticed the men were looking at me sideways. These dolphins never come around boats, they told me. (For good reason—I bet local people have tried hunting them in the past.) So, obviously, I have arcane powers. I'm probably a witch! At the salmon pens they handed me out of the boat onto the floating walkway very respectfully, but they whispered to the other men working at the pens as I passed.

Well, maybe they're right. This witch *can* read the future: I bet they tried the pipe business again after I left, and I bet it didn't work. Having satisfied their curiosity, the cute little dolphins wouldn't respond the second time around. That, too, would have been easily explained by the crew—when I left the fish farm, I took my magic powers with me.

There's considerable benefit in being able to bewitch your research subjects. When I got home I contributed the photos to another scientist's monograph on the genus *Cephalorhynchus*. I also sent a full set of color prints to Ken Norris. End of discussion.

Clicker vs. Traditional Training: What's the Real Difference?

It pays to understand your subjects—just
engaging their curiosity can give you access and some control—
but to really get to know them and train them,
you need something else.

I was talking on the phone recently to a magazine writer—a nice guy who used to be a working cowboy. He spends a lot of time thinking about kind versus cruel animal training.

"So, is that the real difference between clicker training and traditional training?" he asked. "Is it that one is more positive than the other?"

Not really. And, it's not just the food. It's not just the absence of punishment. It's not even the marker, the click itself. All of those are powerful tools for making this training technology work for you. But the real difference is this:

In traditional training, animals learn what to do and what to avoid around people from the reactions of people. It's the same way animals learn what to do around other animals in the wild, from the reactions of other animals.

In *our* kind of training, animals learn how to find food, increase their skills, and discover new ways to have fun the same way they learn what to do in the environment—from exploring the world itself.

We are tapping into what noted Washington State University neuroscientist Jaak Panksepp dubbed the SEEKING system—the part of the brain (the hypothalamus) that governs the urge all living beings have to explore their environment and get good things from it. The SEEKING system has absolutely nothing to do with dominance theories or any of the faux-science around avoidance- and punishment-based training. It's a better way to reach the animal's mind. And it's better because it's *fun*—for both the trainer and the learner. Panksepp discovered that rats laugh (in high-frequency tones) when they're playing. This makes evolutionary sense, as I recount in Chapter 10 of *Reaching the Animal Mind*:

> *The urge to seek and explore needs to function not just when you are in need of food or warmth or shelter, but when you're feeling quite relaxed and happy already. That's when you have the energy and the desire to go exploring, so exploring needs to be reinforcing in itself, or we wouldn't get up and do it for "no good reason."*

> *We hunters and gatherers are totally familiar with this experience. Of all the larger mammals, none of the others come anywhere near us in our ability to find all kinds of new things to do in new environments. I get that SEEKING-system sense of excitement from window-shopping, traveling, snorkeling, bird-watching, crossword puzzles, computer games, turning on the evening news, and playing Monopoly with my cutthroat grandsons. Clicker training is so successful—for us human animals as well—precisely because it engages just that part of the brain we all love to use and get to use too little: our inner happy explorer, our SEEKING system.*

Too Many Butterflies

*When she was a child, Karen's SEEKING system was
alive and well. For a budding young scientist, however, being
challenged about why she was doing what she was doing
stopped her cold—but only temporarily. She's been
following her curiosity ever since.*

From about age nine through thirteen I collected and studied butterflies. My interest was intense; I might well have grown up to be an entomologist. Each summer I visited my father and stepmother, wherever they happened to be; one summer that meant going to Rushford, a little town in upper New York State, to stay with my father Phil, my stepmother Ricky whom I adored, and Ricky's mother, Jennie Ballard. Mrs. Ballard was very critical—a mean woman, my aunt Verona once said—but I don't think she was mean, just outspoken. I was 12, and at the height of my powers as a butterfly collector. I was always out in the Ballard gardens and fields with my net and killing jar (a Kleenex soaked with carbon tetrachloride was the agent that instantly suffocated any insect dropped into the jar).

Little brown butterflies, the wood nymphs, were abundant in the hay meadows. I noticed something odd about them. They all had *ocelli*—circular patterns or eyespots—on the underside of the wings, but in different numbers and arrangements. Some wings had one ocellus, some had four; some spots were big, some little, and so on. I wanted to see if there was any kind of pattern or consistency to this irregularity; it didn't seem right to be completely random, but it was certainly striking.

I didn't have room to mount all the wood nymphs, wings spread, so I could compare them, nor did I have any way to carry them, mounted, back to my mother's house in Ivoryton, Connecticut, after my visit. I knew how to protect them, though, dried with wings folded upward in the normal at-rest position, in little triangular waxed-paper envelopes the butterfly books had shown me how to make. So I collected and killed about 40 wood nymphs, let them dry in their natural state with wings folded, made little envelopes for them, and packed them in a Mason jar to take them safely back to Ivoryton. My intention was to then moisten them as instructed in the books, mount them all on one piece of beaverboard upside down so the undersides of the wings were showing, and have a good look at those ocelli.

Mrs. Ballard stated her disapproval publically, at the dinner table. Her daughter Emmaline (Ricky's sister) had collected butterflies as a child, but she just took one of each kind; she didn't kill dozens! I was powerless to argue; I just shut up. My grandson Max Leabo, finding himself in that situation at age 12, would have spoken right up that this was a research project and explained at great length about the curious variations in the eyespots, but I didn't have that kind of courage. I knew what I was doing and thought of myself as a scientist in the making, but I didn't know how to explain. Nor did I have any justification to go with an explanation—no school project, for example. And now I'd been rebuked and ridiculed for it. So I said nothing.

I packed up my butterflies and went back to Ivoryton. Of course, I did not find the time or energy in the rest of the summer to soften and mount all those butterflies. I left the jar on a shelf in the pantry. Some years later, when I was in college, my mother Sally asked me if it would be all right to throw them out. Yes, I had to agree that would be okay.

Recently my author/scientist friend Alexandra Horowitz gave me a wonderful book, James Elkins' *How to Use your Eyes*. There's chapter in it on looking at butterfly wings. It turns out that both pairs of wings on any given butterfly are identical. Alas, it had not occurred to me at age 12 to look and see if left and right sides were alike! *That* was an oversight; a better observer than I would have checked that out. I would not have had to mount my wood nymphs spread out; just one side of each dried, wings-folded-upward butterfly would have served. Heck, I could have laid them all out on wax paper on Mrs. Ballard's dining room table, grouped by pattern similarities or differences, right there in Rushford. Phil and Ricky, and probably Jennie Ballard too, would have found that interesting.

So the question remained: are there patterns to the variations? And why? I wondered if anyone else did anything about it. Perhaps I should ask Elkins. He's easy enough to reach; he's a professor at the Art Institute in Chicago. And then, looking for a photograph of my little brown butterflies online, I discovered that my wood nymphs belong to a group of butterflies that are outstanding for their curiously variable ocelli. Whew, I am not alone after all.

It's really the curse of the naturalist: trying to explain what you're doing and why. Darwin probably had trouble explaining to visitors the many years he spent preoccupied with gall wasps. Even now, I repeatedly get asked the old traditional question, by friends and colleagues, by my literary agent, (not by family, my family is used to me): "Why are you doing *That*? Isn't *That* a waste of time? You could be making more money doing something really important! No one's interested in *That*." Oh well. Like Kipling's Elephant's Child, I might get thumped for my insatiable curiosity, but I'm not going to stop.

Playtime

The SEEKING system is not all about the serious stuff: survival
and reproduction. Sometimes it takes animals down a decidedly
different path, one that we can all relate to.

I've always been interested in play. Science doesn't explain it very well, or
it's explained as something young animals do to practice future skills. But
that definition doesn't cover every kind of play, and it doesn't explain why
it's so much fun, so reinforcing in itself.

Fish games

Helix Fairweather, a Karen Pryor Academy and ClickerExpo faculty
member, is training a fish and posting videos of her progress on YouTube.
She suspects her fish, Cartman, is playing when he bumps a springy plant
leaf over and over. I think she's probably right.

You may remember the cichlid fish I taught to swim through a hoop
years ago. When my fish grew big and had a big tank to live in, he liked to
play. If visiting children put their noses against the glass, the fish put his
nose against theirs on the other side. If they put their hands on the glass,
he would put his nose to their hands, too. The children could get him to
swim from one end of the tank to the other by putting their hands on one
end of the tank and then the other. There was no click or treat for this; as
far as I can see, he just did it for fun.

How to play with an octopus or a rhino

Some years ago, *Natural History* magazine ran an article about the giant
Pacific octopus. One aquarium gave their big octopus a plastic childproof
pill bottle with a treat inside to see if the octopus could get the lid off

(octopi are good at that sort of thing). Since the bottle had air in it, it was buoyant.

Here's what the octopus actually did with the bottle: He held it up to the jet of clean water coming into his tank, let it go, watched it tumble across the tank, caught it, and put it back in the jet again. He did this over and over—he was playing! (The article made me feel that it might be really fun to train one of those guys!)

In 2009, I visited the Dallas Zoo's excellent rhinoceros collection. The keeper introduced me to her special favorite rhino, a half-grown male who was born at the zoo. The keeper showed me some of the rhino's husbandry behaviors. I enjoyed the experience of personally moving him about with the target pole. The reinforcer was bits of banana, which he took from my hand with gentleness and skill.

For this young rhino, however, the big reward at the end of a session was not food, but a game with the keeper. The keeper left the target, ran the length of her side of the paddock fence, and stopped. *She* was the target. The rhino took off at a run down his side of the paddock, screeching to a halt just as he got abreast of the keeper. She turned and ran up the fence. He chased her back, tail in the air and little eyes sparkling. Of course, this is what young rhinos do! They play "Charge!"

People play, too. I think I'll go dig holes in my garden.

Clicker Training as Play

Clicker training is exploration. You can think of it as a form of
play between two individuals, both communicating with each other. Here's
an example, involving Karen's surfing ponies and their young trainers.
Clicker training also offers the shell-shocked—like some captive
ibises—a glimpse of order and predictability. With the
world a safer place, what happens? They start playing.

Ah, summer! For weeks I haven't been able to get anyone on the phone, businesses don't answer their e-mail, professors are unreachable, my family is camping on the beach, and trainers I need to talk to are tracking down their ancestors in Iceland or bird watching in Belize or going fishing.

Everyone is playing. Play is a highly important part of life. I think it's also a highly important part of clicker training. No, I don't mean as a reward—following the click with a game of tug, say, rather than a treat. That's okay in its place; but that's not what I mean.

I think clicker training, properly done, *is* a form of play; and I said so in print more than 20 years ago. "The trainer, in addition to being largely limited to positive reinforcement, is interacting with the animal; he or she can see the animal, the animal can see him or her, and both can introduce changes in the training process, at will. It is a situation both rigorous [i.e., governed by mutually accepted rules] and admitting of spontaneity: a game."

If clicker training is a game, it's one the animals often win, and like any good game, they love it. And it resembles play in many ways. We engage in clicker training in bouts, with rest periods and other activities in between, as with play. Like play, it calls for ingenuity and initiative from both players. Like play, it's reinforcing. Like play, it requires some learning. Like play, it teaches, painlessly. You can use clicker training to teach play itself, such as tug-o-war for timid or inexperienced dogs. Or, you can use a dog's clicker skills to teach entirely new games.

Horseplay

When I lived in Hawaii I raised Welsh ponies. Each year, a new group of two-year-olds came over from the main herd on Maui and were stabled at a farm a bit inland in Waimanalo Valley, a couple of miles from my rental house on the beach. I had enlisted a group of five children, approximately 10 to 14 years old, to train each year's crop. Two or three days a week the "pony children" were carpooled to the farm after school, where they each took care of and trained a particular pony. Sometimes I was with them, but sometimes not, since, due to the clicker training approach, the ponies were safe and friendly, and the children knew what to do.

One day I came home to my house on the beach and was startled to find the yard full of ponies, each with a child on its back. They had no business being here. To get here from the farm they would have had to ride down the side of a busy, narrow highway, full of huge and speeding trucks. This was absolutely forbidden. Not only that, the children were riding bareback, and they didn't even have bridles, just halters and lead ropes. Oh my goodness, how dangerous! I started to scold, but the children interrupted, "Mrs. Pryor, Mrs. Pryor, come see the ponies surfing!" And they all went cantering around the side of the house onto the beach.

Well, I had to shut up and go look, didn't I?

There was, as usual, a gentle shore break, four-foot waves rolling in steadily on the long, straight Waimanalo beach. Each of these young ponies waded right out into the water and headed to sea, child aboard; obviously this was not the first time they'd done this! As the wave rolled in, the white water foamed right over child and pony, so they completely disappeared, reappearing on the far side of the break. Out there, it was too deep for the ponies to touch bottom. The ponies began dogpaddling back and forth parallel to the beach. All heads were turned out to sea. Ponies and children both, they were all watching for a good wave.

When a bigger wave came, the ponies turned toward the shore and began cantering; the wave whooshed them toward the beach and when they felt bottom they cantered up onto the sand. The children jumped off. The ponies dropped onto the sand and had a great roll, then stood up, shook themselves, and waited. The children jumped on the now sand-coated ponies and back they went into the water, out past the break, to catch another wave.

One old mare, Gaylight, would have nothing to do with this sport and had simply come along in spectator capacity, like me. But the rest of the ponies obviously loved it. Summer fun for horses. And why were the ponies in halters only? The children had left bridles and saddles behind so they wouldn't get them wet.

Yes, I put a stop to it; the highway risk was too great. But it sure was fun, and I'm glad I saw it myself.

Bird play—as an unexpected benefit

Elsa Mark, a bird keeper at the Philadelphia Zoo, was faced with a major problem. The zoo hosts a colony of one of the most endangered birds in the world, the Northern Bald Ibis, *Geronticus eremita,* a native of Morocco and Turkey. By nature, these are nervous birds, easily upset, distrusting of humans. No matter how long they had been in captivity, they never seemed to get used to people. They panicked and went flapping into the walls when keepers were working in the area. They had a nasty skin problem of unknown origin, causing them to peck at themselves and each other. Any medical procedure, even simply weighing a bird, was a scene. Writing in *Wellspring,* the journal of the Animal Behavior Management Alliance, Elsa Mark commented, "Needless to say, a typical day of caring for these birds can be stressful for the birds as well as the staff."

The zoo arranged for a training workshop by Steve Martin and his staff from Natural Encounters, Inc., a leading resource in operant conditioning, training, and management for birds and other exotic species. The birds learned to eat small amounts often. They learned a marker signal, the word "good." *That* made it clear to them that "arrival of keeper" meant "arrival of food," something they'd apparently failed to figure out in the past (since they were previously fed only once a day, and spent that time flapping about in panic). Once they acquired these basic insights, training proceeded rapidly. The birds learned to stand on a green mat, one bird per mat, for "good" and a treat. They learned to get on the scales. One by one, they learned to go in a carrying crate on cue, and stay there while the door was shut.

Eight new ibises arrived and had to be taught, too. A new exhibit was built and the birds were moved. The incidence of startling and panic decreased to zero for most birds; stress indicators were dwindling in all the birds. The new birds took their cues from the old birds, and calmed down right away. And on September 16, 2003, just a month after the eight new birds had been introduced, one of the old birds was observed, for the first time ever at that zoo, playing with a leaf.

The next day, two birds were involved. The two birds would steal the leaf from each other and walk around the exhibit clacking it in their beaks. Within days, lots of birds were playing with leaves, including birds from the new group.

The keepers rightly thought that this was a fine thing, and supplied the birds with spoons, plastic discs, and a variety of wooden and leather toys and puzzles made for parrots. Yes, the birds played with them all, and even in the presence of the keepers.

The change in these rare birds has been profound and permanent. The birds are much healthier and, I can't help but believe, happier. The keepers enjoy them. Visitors to the zoo love to watch the training and the play. Not only are the birds no longer afraid of the keepers, some of them take an interest in the keepers' work and follow them around, actually getting in the way. And play, in and of itself, turned out to be a great indicator of the benefits of operant training as well as something no one *ever* anticipated seeing in the Northern Bald Ibis. The potential for saving this species has improved; thanks to playing with a leaf.

Summer fun for the planet.

101 Things to Do with a Polar Bear

*The animal world, Karen points out to us, abounds in
creativity—it's all part of fun and games.*

I recently posted some reports from the zoo world about using the
"creative" game—training an animal to think up its own behaviors—with
a gorilla ("101 Things to Do with a Gorilla"). People sometimes think of
the training we do as something artificial. Why would you want to demean
the animal by making it play games? Never mind if the animal loves the
game! Furthermore, to some people the "101 Things to Do with a Box"
game seems particularly confusing and problematic when applied to dogs.
How does the dog "know" what to do if you don't tell it? Why would you
want it to just guess?

This is a misconception deriving from conventional, coercive training,
in which the goal is to control the animal's actions (as long as you are
around): to make the animal stop doing the wrong things, and only do
what you tell it. Modern training relies much more on the animal discovering what works. Exploring and innovating is part of the way animals learn
naturally, on their own. It's very much related to play. That's one of the
things that makes clicker training so much fun, for the animal and for the
person, too.

Here's an example, from a baby polar bear…

On July 8, my grandson Nathaniel's sixth birthday, he commandeered
his Mom, his Grandma (me), his older brothers, and his friend Taylor, a
six-year-old girl who lives on his street, for a trip to the zoo. Roger Williams
Zoo, in Providence, Rhode Island, is a small but very pretty and well-
thought-out zoo. We liked the giraffes and the elephants and the Thorny
Devil (an Indonesian lizard), but the hit of the day was indubitably the

baby female polar bear (2 ½ years old, 300 pounds) playing in her pool while her 600-pound mother snoozed nearby.

When we got there the baby bear was in the water, up against the underwater viewing wall of her pool, bewitching a bunch of children. Nattie and Taylor sat on the floor, noses and hands on the glass. The bear turned herself upside down in the water and put her nose against Nattie's nose. Then she put her nose against Taylor's hand. Three times. Taylor put both hands flat on the glass over her head. The bear turned itself right side up and put her own big paws flat against Taylor's hands, on the other side of the glass. This bear was playing 101 Things to Do with an Audience!

Every so often, perhaps three times a minute, the bear had to swim up to get a breath. Like a click and treat, that provided a sort of break in the action, and often, when she dove down again she did something new. Like what? Well, she could turn and wiggle in any direction. She could see the children very well, and followed them, poked at them, and made faces at them (scary-bear faces). She tried blowing different amounts of bubbles. She made noises. Once she hung from the pool ledge by her back feet for a while, her own back against the glass, looking over her shoulder.

Gale, Nat's mother, crouched down to the glass next to the children and the bear rushed her, darting at her sufficiently aggressively to make her jump back involuntarily. "That's a great behavior! Get Mom!" I joked with Nattie. The bear liked it too, because she went on a riff of play attacks. She rushed at a wadded-up shirt in a child's lap. Other kids immediately produced wadded-up garments, and the bear darted at those (parents began to worry about whether the glass was really thick enough). One boy unfolded a zoo map and spread it on the glass. Wham, the bear feigned an attack on that, this time with an audible underwater growl.

I think my favorite behavior of all was one I observed later, from the surface. The bear was seeing how long she could stay upside down under water—while holding her left hind foot in the air.

All in all, from my hasty notes made that night, this big baby bear came up with around 60 behaviors in about 10 minutes, about half of which involved getting the kids doing something, too.

And it made me think. We often say that with clicker training we see the best side of the animal. We like the 101 Things game because it brings out the animal's mental capacities. And here was a wild animal, captive-born but not trained (this zoo does not do that, I'm told), using these innate capacities on her own. Part of being a successful polar bear must include developing initiative, imagination, flexibility of mind and body—and a certain sense of humor. That's what this baby was working on.

I think the "game"—and the communicative aspect of our clicker work—is far from artificial. It exemplifies Mother Nature at her best. So much of the behavioral world divides what animals do into "natural'" behavior, such as dominance, fear and aggression, and "behavior modification" or training. In truth, I think it's a continuum, with a lot of overlap. Let's bring these diverging views together; let us talk about the whole animal, brain included—it's the separation that's artificial.

The View from the Bridge

A t its heart, clicker training is a form of play. It is also a conversation. It's very much a two-way street. An observant trainer can initiate and guide a conversation, for instance, that helps a naïve puppy cope successfully with the overwhelming environment of a TV studio. With clicker-savvy animals, however, the give and take is more robust. They keep you on your toes—or even clicker-train you themselves!

Karen holds that the more you can marry ethology (observing innate behavior) with behaviorism (studying learned behavior), the better conversations you can have. She shows you how. The conversations are nonverbal but she puts metaphoric words to them, and you get an insider's look at how her mind works.

Arizona Sonora Desert Museum Raptor Specialist Wally Hestermann works with one of the Museum's Harris's Hawks (see page 31). Photo: Jay Pierstorff

The Ethology of Clicker Training

Animal behavior is not a unified science: The
naturalist types study innate behavior, and the lab types
study learning. Karen's aim is to integrate the fields
so we end up understanding, appreciating—
and respecting—the whole animal.

Since the 1940s or so, there have arisen two completely different kinds of animal behavior science. One discipline focuses on what animals do in the natural environment: the innate behavior of the species. Innate behavior—hunting, foraging, reproduction, dominance hierarchies, sexual displays, and so on—is a product of evolution and is largely governed by genetics. The field is called *ethology*, or, simply, animal behavior. Its best-known founder was Konrad Lorenz. There are departments of animal behavior in universities all over the US and Europe. You'll find them in the biology buildings.

The other group of scientists studies acquired behavior: how animals and people learn. The field is called *behavior analysis*, or simply behaviorism. Its best-known founder was B.F. Skinner. There are departments of behavior analysis in universities all over the US and at universities in South America, Europe, and Asia. You'll usually find them in the psychology building.

I have been interested in the natural behavior of animals all my life. I was one of those children who watch animals, keep aquariums, collect butterflies and moths, and know the name and habits of every bird, tree, and wildflower in her neighborhood: a naturalist. In college, I discovered Konrad Lorenz's books and found a name for my natural bent: ethology. Watching animals to see what they do. Then, a few years later, I took a job as head dolphin trainer at Sea Life Park, the Hawaiian oceanarium complex pioneered by my first husband Tap Pryor. That's when I discovered operant conditioning and the work of B.F. Skinner.

I was fascinated, not so much by the dolphins themselves as by these simple yet powerful laws of learning. I quickly found out that these new tools enabled me to understand dolphins—and sea birds and fish and wild pigs and anything else I decided to train—on a new level. I could learn a lot by watching my animals to see what they did when undisturbed. I could learn even more, however, from our reinforcement-based interactions, in which the animal and I were equal players. We communicated in both directions: I with my carefully timed reinforcers, the animal with species-specific signals and emotional responses directed specifically, and with intent, to me.

And they made themselves perfectly clear. Want an example? Guess what a dolphin does when you accidentally frustrate it by suddenly *not* reinforcing behavior you used to pay for, a mistake I sometimes made in my early training days. The dolphin gets mad, of course. Behaviorists have a name for it: extinction-induced aggression.

In that circumstance the dolphin sometimes breaches. Breaching is a whaling term for the behavior of leaping out of the water and coming down sideways, making a big, noisy splash. Now if my frustrated dolphin breaches, and the splash from that breach happens to soak me from head to toe, and if that dolphin then pops its head out and looks at me with a glint in its eye as I'm wringing the sea water out of my hair, I have just learned something. Breaching can be a message about an emotional state, which a human might express as: "I'm thoroughly fed up with you!"

In my view, if you look at just the innate behavior of animals, you are looking at just a part of what animals (and people) do. And if you look at just the acquired behavior, you are looking at another part of what animals (and people) do. To really understand what's going on, you need to be able to use both parts at once: to observe accurately, not with blinders on—that's the ethology part—and to use the principles of reinforcement, without aversives or coercion, to facilitate learning with confidence and joy—that's the behaviorism part.

Konrad Lorenz visited Sea Life Park and watched the dolphins and the training with interest. When I later wrote a book about my dolphin days, *Lads Before the Wind,* Konrad kindly contributed a foreword in which he perceived the ways in which the two behavioral sciences can interact: "Karen is one ethologist who uses the whole arsenal of methods devised by the behaviorist school, not only to study the contingencies of reinforcement, but as a tool to gain knowledge about the animal as a whole." Exactly. But why should I be the only one?

Opposite ends of a bridge

In the years during and since my dolphin training days I have been a participant in both scientific fields. I belong to both societies (the Animal Behavior Society [ABS] and the Association for Behavior Analysis [ABA]). I publish scientific papers in both kinds of journals. It's not common; I personally know only three other people who belong to both ABS and ABA. Both branches of science have a lot to tell us, but the people engaged in them almost never mix.

I used to think of myself as standing perpetually on a bridge, with a foot in each camp. I used to expend a lot of time trying to talk psychologists into understanding or at least coming to watch what we were learning about the animals with their science. No luck. No luck in the other direction, either: the behavioral biologists were not much interested in training or reinforcement. I once participated in a Navy-sponsored conference on dolphin cognition, consisting of about 20 hand-picked famous scientists and me. After we'd listened to one long story after another about how some dolphin had done some amazing thing that demonstrated "cognition," I finally spoke up and described how one trains that kind of behavior. The ability to problem-solve is an outcome of the reinforcement contingencies used in dolphin training, I pointed out. Instead of everyone saying, "Oh! Right! Of course! Now how can we use that," there was a long silence. Then a famous brain scientist said "That's Skinner stuff, Karen. That's so out of date, you'll never get a grant with that!" Of course, I didn't need or want grants so it was an ineffective threat, but I got the message: shut up, Karen.

In any case, marine mammal trainers often speak about the work and the animals in ways that ruffle the feathers of academicians. Once at SeaWorld, the head trainer and I were watching a junior trainer working with a killer whale. The killer whale was lolling sidewise some distance away, keeping one eye on the inexperienced trainer as he tried to make the whale jump. The head trainer laughed, and spoke for the whale: "You and what army?"

Putting words in an animal's mouth like this is often labeled anthropomorphism; it is not. The trainer is illustrating a behavioral event, often but not always in a training interaction, by using the language a human might use if the human were in that particular situation and emotional state. I call it trainer metaphor, and to me it is an indication that the speaker is using both sciences.

The science that pervades the dog world these days is ethology: the genetic or biological approach to behavior. Trained in ethology, applied animal behaviorists use this science therapeutically. One might correctly identify, let us say, submissive urination in a pet and suggest ways the owner can behave to mitigate the fearfulness. Many dog trainers and instructors have incorporated concepts from ethology into their practices or writing. Pet owners often have at least a loose acquaintance with ethological terms such as "alpha animal," "dominance," "territory," and "aggression."

But a general awareness of animal behavior doesn't mean people read the animals accurately. I am amazed at how often people fail to recognize canine signals of simple fatigue, much less signs of real stress. Pet owners interpret threats as play and play as threats. Traditional trainers and dog sports competitors ignore that chronically worried expression—"Oh dear, now what am I supposed to do?"—that I call the "crossover look." Charismatic dog trainers on television borrow phrases such as "dominance theory" to justify terrorizing someone's yappy Yorkie into never barking again. As the cowed dog's expression of bewilderment turns rapidly into misery or fear, its owners, the audience, and the TV producers who created the show unquestioningly accept the trainer's methods and explanations. And, when a once overly bouncy dog is now hiding under the furniture with its tail between its legs, that's seen as an improvement. Clicker trainers can't bear to watch.

Using both sciences

You don't need two PhDs to use both sciences; that ability seems to be a natural outcome of the clicker experience. At our ClickerExpo in Minneapolis in November, 2005, we introduced a new feature: Learning Labs. These were sessions where people with dogs could try out what they had just learned in a lecture about some aspect of operant conditioning, and other people could watch. I taught two Learning Labs and visited several others. I was thrilled to see that many of the spectators were visibly relishing both the operant conditioning procedures they were watching and the animal behavior they were seeing.

During an exercise on transferring a cue from the voice to an object, a trainer was timing her cues wrong, so they didn't make sense. The dog began to bark at her. In trainer metaphor, the dog was saying, "Tell me what you mean, darn it, I don't understand!" And several spectators smiled kindly. They were not frowning, thinking, "That dog is barking. What a nuisance! It shouldn't be allowed." They understood what generated the protest, and they were both sympathetic and amused.

In a shaping exercise, I chose three people and their dogs to develop a behavior involving a box. One dog leaped over its box; another quickly learned to step up with its front paws on the box. The third owner was shaping her golden retriever to put both front paws inside the box. Now the first two teams sat down. The room fell silent while we all watched the dog figure it out. There were gasps, laughter, and nods, not when the trainer made a smart move, but when the dog did. When the dog's tail started to swing ("Yeah, I think I've got it!"), applause pattered around the room—not for the finished behavior, but for the dog's awareness of progress. People were truly seeing what was going on with the dog. I was watching people exhibiting dual perceptual skills. Seeing the animal whole. Enjoying the view from the bridge.

I'm beginning to think the clicker training movement might, in coming years, make a real difference to both of the mother sciences. And I can tell you, my fellow trainers, it feels great not to be out on that bridge all alone!

Clicking Raptors in the Saguaro

Skillfully combining knowledge of innate behavior with training
gives not only scientists but tourists insights into the complexity of
Harris's Hawk society. Everybody benefits, including the hawks.

We're standing on a gently sloping foothill with Tucson's jagged volcanic peaks behind us, looking across the vast, flat Avra valley far below. The hills beyond that valley are in Mexico. The desert sky is a brilliant, piercing blue, filling the eyes with light. The mild warmth of the winter sun is welcome. This is Saguaro National Park. Giant saguaro cacti, their upraised arms frozen in mid-gesture, dot the rolling terrain around us and march up the slopes of the mountains behind us like some strange army. Between and around them, desert-adapted plants thinly coat the sandy ground: mesquite and brittlebush, palo verde, prickly pear, and ocotillo. To my temperate-climate eyes the plant life might as well be from another planet.

Someone points out a hawk, sitting on the top of a saguaro perhaps 50 yards away. Just as I spot the bird, it lofts itself into the air and flies closer to us, landing on the top of another saguaro nearby. We are looking at the bird, and the bird is having a look at us. It's a Harris's Hawk, and it's Harris's Hawks we've come to see.

Most hawks are speckled with light and dark brown in various patterns that change with age, region, and gene pool. That can make identification difficult for the amateur birder like me. Harris's Hawks, conveniently, are unmistakable: a solid, dark brown all over, with copper-colored shoulders and a handsome tail with dark brown and white bands. They fan the tail when they land, wag it when they are excited, and spread it often in their acrobatic aerial maneuvers. It's the Harris's tribal flag, identifying them even at considerable distance.

Harris's Hawks are unique in another way. Most raptors are solitary as adults; these hawks live in family groups. See one, look for more: sure enough, here comes another, a larger bird, therefore a female, landing on the top of another saguaro. Now comes a third, and then a fourth. The last one barges onto the saguaro top the first bird is sitting on. "I deserve to perch there!" The bird already in place knocks the newcomer away with a well-timed kick. Then all four birds hop into the air and circle, trying out different perches here and there—musical chairs on the saguaros.

Harris's Hawks are territorial. Each family jointly defends a piece of land that, in good Harris's Hawk country, will be about a mile square, what ranchers call a section. They want a water supply; an abundance of rabbits, rodents, and other prey; and long sightlines to sport interlopers, such as other hawks, and dangers, such as coyotes and eagles. We are in the middle of this group's highly desirable territory.

We are also in the middle of the grounds of the Arizona-Sonora Desert Museum, a unique collection of living plants and animals from the terrain around us. Some of the museum staff and volunteers attended Clicker-Expo in Tucson this past weekend. They invited the ClickerExpo staff to visit their unique facility, and several of us stayed over an extra day to do so. We're not the only ones here to see the Harris's Hawks. About 60 other museum visitors have joined us at the appointed spot, a widening in the path through the desert. A visiting camera crew is crouched amid the prickly pears photographing the birds, and a museum docent with a portable loudspeaker is telling us about Harris's Hawk behavior and natural history.

Now the female hawk departs purposefully toward the east, soaring just above the bushes, and landing with her wings outspread on a sandy patch a hundred feet away. She is mostly out of sight, but I can see she's mantling, or holding her wings open, and bending her head to the ground. Her breeding partner joins her, and, briefly, the other two birds. We couldn't see what took them to that sandy spot, but I suspected it was breakfast. The narrator on the loudspeaker had told us that Harris's Hawks feed on rabbits, birds, ground squirrels, and other small rodents. She tactfully does not mention that the birds might be hunting—and killing and eating—some small animal at this very moment.

Now the hawks all rise and come back to land together in the bare branches of a nearby tree. For a moment all four are in one place. The senior pair actually sit side by side, amicably touching shoulders.

This wonderful chance to watch some unusual bird behavior in the wild is not an accident; it's a current feature of the museum's educational offerings, called the Raptor Free Flight. Every day, at 10:30 a.m. and 1:30 p.m., the museum's animal behaviorist, Sue Tygielski, PhD, and her team display for visitors the behavior and activities of various, free-flying desert birds. The birds are the focus; you will hardly notice the trainers. And at 1:30 p.m., it is the Harris's Hawks.

Come back, but don't come down: An advanced exercise in cueing

Right now Sue Tygielski and Marta Hernandes are crouching more or less out of sight among the bushes, communicating with each other by radio headsets. They are not interfering with the birds, but they are, from time to time and very unobtrusively, reinforcing them. Sue stands up and looks across the crowd. One of the hawks flies toward her. The bird coasts in on wide-spread wings right over our heads, almost within reach; we hear the feathers rustle and feel the wind of its passing. The crowd exclaims. Then another bird soars right over us. It will happen several times, and it is an exciting experience each time.

These birds don't seem to mind being around people at all. Several years ago, when the female was just under a year old, the two males made such pests of themselves during the breeding season that she took to landing on the path and standing around among the spectators, just to get some peace and quiet! Now she coasts over spectators' heads and lands, not on the path, but in a nearby saguaro. This happened not through aversives of any kind but by selectively reinforcing incompatible behavior: Join the spectators, nothing happens. Sit on a cactus, get a treat.

As another bird soars over the heads of the crowd, Sue "clicks" the bird in midair, not with a sound but with a cue for another behavior. She has cued the bird to land on a specific dead palo verde tree nearby. The cue is her glance; the hawks have learned to go where the trainer is looking. The bird lands in the indicated spot, and Sue, turning her back to the audience, quietly delivers a treat.

Sue did not reward the gliding bird for coming close to the crowd, but rather for staying far enough away. These birds often hunt stealthily by flying very low to the ground. Left to themselves, Sue tells us later, the birds seem to enjoy using people as an obstacle course, flying through the crowd rather than over it. It frightens some people and holds some risk of injury for both birds and spectators (especially children, who might move unpredictably into the flight path). So the trainers, using cues as the bridging signal or event markers, shape and maintain fly-ins at a suitable height and a brisk speed.

Now all four birds are perched near Sue in a low, bare-branched tree. Sue reinforces the youngest male, still a bit inexperienced, for joining the group. After eating his treat he tries once more to sit where the senior male has chosen to sit. The nerve! The two males have a splendid midair argument, grappling with their bright yellow feet, locking talons, tumbling earthward in a rush of beating wings, and flying apart just before they hit the ground. This dramatic scene takes place right in front of us, not 20 feet away. It's like stepping into an Audubon painting.

The young male stays on the ground for a while, striding about on his long yellow legs. Most raptors are rather awkward on a flat surface, but this fellow looks at the moment amazingly terrestrial, rather like the ground-loving Caracaras I have often seen in Chile. The narrator confirms that impression. Sometimes, she tells us, during a Harris's family hunt, a rabbit being chased simultaneously from several directions at once will take refuge in a dense patch of brush. Then one bird, usually the male with the lowest status (thus the most expendable should a coyote show up) lands on the ground and walks or crawls into the brush to flush the rabbit. Junior here seems highly capable of that feat of daring.

Now, apparently, it is time to end today's Raptor Free Flight. Sue picks up the young male on her glove. At a cue invisible to me, the other three birds loft into the air and sail away, disappearing over the crest of the nearest hill, going home. Sue walks after them with the young bird on her fist (later she explains that he hasn't quite got the hang of the "Come home" cue yet, so he still gets carried home). The narrator announces that one of the trainers will be back shortly, with a hawk in hand, and invites us to stay and ask questions. As we wait for the trainers to return, the ClickerExpo group members all agree that it's been one of the most thrilling animal shows we've ever seen.

But falconers have been training raptors for thousands of years. What makes this so different?

I've talked and written about the schism between the two kinds of animal behaviorists: the behavioral biologists, who study evolutionary aspects of behavior, and the behavior analysts, who study learned or acquired behavior. I have often spoken about feeling as if I've been standing on a bridge all my life, with a foot in each camp, trying to encourage each side to cross over. The two camps don't talk to each other much, and very few practitioners combine the two sciences routinely, using each science to facilitate the other.

But here we have an example of the merging of the two sciences. Bird shows with free-flying birds were pioneered by trainer Steve Martin and are now popular attractions in zoos. In these shows, each hawk, owl, or parrot involved is exhibiting trained behavior solely: often what zoo trainers call

"A to Bs," that is, flying from place A to place B on cue. It is wonderful to see these marvelous zoo creatures flying at all, but they are not just zooming around doing whatever they would do on a normal Monday at home—which is what these Harris's Hawks are doing.

This is indeed a view from the bridge. Dr. Tygielski and her team are using their understanding of the birds' ethology to enable us to have an incredible view of complex social behavior. They are also using highly sophisticated reinforcement techniques to keep the birds around, and to care for and protect the birds—without any of the coercive techniques of traditional falconry. While aversives are not a big factor in falconry—none of the falconers I've met would dream of punishing a bird—food deprivation is pretty important. If you are getting your bird to come back to you solely by showing it food, that bird better be pretty hungry, hungry enough to give up the pleasure of flying free; and that means monitoring weight very closely, sometimes at the delicate point between hungry and starving.

The Desert Museum's free-flying raptors, on the other hand, are kept at normal weight. They are trim, muscular, glossy, and in fabulously good condition, like really healthy wild birds. The trainers don't mind if the animals do some hunting on their own. The birds still want the treats because they "win" them by performing their own learned skills. Earning reinforcers, for these birds, is a thrill, not a necessity. And, like us, they enjoy that experience of success.

For the hawks, as for dolphins that work with humans in the open ocean, there's another great benefit to having some reliable humans on their side: safety. I've seen dolphins that I was responsible for, that were swimming around at liberty under the docks at Sea Life Park, jump back into their pens when a shark showed up. Here, Sue tells me, she sends the birds home at once to the holding facility, called a mews, if dangerous golden eagles appear, or if a flock of ravens comes in sight; ravens are smaller than these hawks but they can and will gang up on them. I am pretty sure the birds appreciate the usefulness of a refuge.

And what, I asked afterward, normally brings the show to an end? (I was thinking as if the birds were dolphins, do you quit before they get tired? Or maybe before they get full?) Not at all. The end of the show depends on human ethology. The trainers judge the crowd's behavior. The museum cannot provide shade—a canvas roof, say—because you wouldn't be able to see the display, much of which is overhead. But the desert sun can be pretty severe. If it's hot and the people look stressed, the birds might be sent home after 15 minutes. If the crowd stays alert and seems to be comfortable, the birds might be out playing musical chairs on the saguaros for 40 minutes or more.

Can you go see this for yourself? Sure.

Would I go again? You bet. There are big events we did not see. For example, Sue regretted that the birds hadn't shown us any soaring, a spectacular sight; apparently there were no thermals or spiraling vertical flows of air to carry them skyward that day. Sometimes another hawk shows up, a redtail, say. Like military jet planes when a strange airplane crosses a border illegally, the Harris's Hawks "escort" the intruder out of their territory; I'd like to see that!

But what really makes me think about keeping Tucson in my future travel plans is this fabulous display of the marriage between natural behavior and operant conditioning: this wonderful view from the bridge.

Bird Talk

*In the hands of an expert, the principles of clicker training
enable a higher level of inter-species communication,
companionship—and freedom.*

It's cold. It's winter. It's boring.

Let's travel, at least in our minds. Let me introduce you to my long-time pen pal, Shanlung, probably one of the best bird clicker trainers on the planet.

A world-class bird clicker trainer

Shanlung is not his real name; this pen name means Mountain Dragon. Shanlung is an engineer who designs and builds huge projects—bridges, dams, harbors—in remote parts of the world. During assignments, he often brings along his wife, his cats, and his birds. They do quite a lot of truly frightening sightseeing (frightening by my standards, hating heights) on motorcycle. Always with a bird.

Most of Shanlung's birds have been parrots, trained to free-fly, an elaborate procedure starting with the bird on a harness and safety line. If you prowl through his journal, you will see him flying one or another bird around Tibetan monasteries, Arab desert villages, and the abandoned wadis and ruins where he and his wife like to picnic on his days off.

Reading the bird-ish mind

Between assignments, Shanlung and his wife live in Singapore. In 2012, Shanlung was experimenting with some of the native songbirds Singaporeans keep in cages as pets. (These birds are usually released in appropriate territory in due course.) Shanlung reports his findings through his live journal, findings that range from his careful work on diet design to observing, recording, and responding to social behavior directed toward the trainer.

The interesting thing to me, and to Shanlung, I think, is not just the training successes—getting the bird to fly free around the apartment, come on cue, target, and so on. It is discovering, through the clicker interactions, how that species of bird thinks, and what is appropriate and interesting for it.

Clicker training doesn't just take away the fear and give you control. That's the utility, but that's not the key benefit. More exciting is the way it builds real communication with the animal, whatever the species. Trainers become, to the animal, a sort of weird species-mate: *"I understand what you mean, and I presume you understand what I mean, too."*

The animal is not "trying" to communicate; it *is* communicating, by giving you, the trainer, the same emotional information it would give another animal.

Go adventuring with Shanlung

An excerpt from Shanlung's journal (http://shanlung.livejournal.com/135250.html) from 2012 starts out with a complaint about his new cat (who has him perfectly trained, as you will see). What follows is a running account of his current bird project—Jackie, a wild-caught adult mynah that is proving to be a very strong-minded individual with a tempestuous, emotional nature. Again, as you will see, if you follow the link to his journal.

A lot of bird specialists (zoo people and so on) follow Shanlung for his aviculture wisdom. I follow him for training sagas and the careful thinking he puts into each interaction. I also follow him for the jokes, such as the two brief theater reviews in the blog excerpt cited here.

I hope you enjoy "meeting" Shanlung. Be sure to browse the photography in his previous journal entries—awesome!

Bad Bob

Through clicker interactions with two horses in Ireland,
Karen uncovers how two horses think—and feel—and
what is appropriate and interesting for each of them.

August is a time to relax. This month I just want to share with you a personal experience: one wonderful summer day of clicker training horses. Beautiful Irish horses. I just had to tell you everything, and it turned out to be all one story. So print it out and take it to the beach.

"It's the pinnacle of a sparkling career," my friend Joan said, "working in a horse barn in Ireland." It does seem a little coals-to-Newcastle, shipping an American grandmother to Ireland to teach people to train horses, since the Irish have been training horses as well as or better than anyone else for a thousand years or so. But clicker training was a new idea to them, and they'd asked me to come.

TAGteacher Theresa McKeon and I were in Ireland at the request of the Festina Lente Foundation, which has premises just outside Dublin. They teach people with disabilities to work with horses and in horticulture, and to become employable. They also run a large riding school, giving lessons to children and adults. We had spent several days participating in a Festina Lente conference on TAGteach: using the marker signal to teach people. Festina Lente staffers had been using TAGteaching with considerable success with their human students and presented some wonderful case studies at the conference. Theresa and I gave talks and workshops on TAGteaching, clicker training, and learning in general.

The day after the conference, we went out from our hotel in Dublin to the foundation's riding school. Theresa was scheduled to work on TAGteaching with the riding coaches and their classes in the arena all day. I'd agreed to demo some clicker training with horses.

I did bring some equestrian experience to the job. Many years earlier I had owned a herd of Welsh ponies and trained dozens of them, many from birth onward. Of course I didn't do this all by myself. I had a group of volunteer children, ages 10 to 14, who took a mix of old broodmares and new, untrained two-year-old ponies, taught them, and rode them for a year or so, watched the youngsters go to new homes as kid-proof, ride-and-drive-safe ponies, and then started again with another group. I taught the kids how to use positive reinforcement and a marker signal as we did with dolphins. Then I thought up assignments, and they did all the work. Easy for me! And the kids called it "playing ponies," as if the whole thing were a game. Which, of course, it was.

The children were great, and we all loved working with the horses. They are beautiful, they are amusing, and, when you train them purely with reinforcement, you find out that they have very interesting minds. It had been years since I'd "played ponies." Now for a day I would once again have a chance to train some horses. My way. I was excited.

Naturally, Jill Carey, director of Festina Lente, wanted me to begin with their biggest problem. It's not the ideal place to start learning clicker training, in my opinion, but people always want the problems solved first, and I accepted that. In this case, the problem was a nine-year-old Connemara pony named Bob. They'd acquired him a couple of months earlier from an old lady who loved him dearly but who was afraid of him. In fact, she was so reluctant to handle him that he'd spent the last few years of his life turned out to pasture, with no company and nothing to do—a very lonely life.

It was clear to see why his owner had banished him. He had already annoyed so many people at Festina Lente that he was now called Bad Bob. Interestingly, he had also annoyed quite a few horses; he apparently had no horse manners, either, and was wearing several dings and nicks as proof.

Clodagh Carey, Jill's sister and the head of the riding school, took me to see Bob. Jill and Clodagh are warm, lively, pretty women, one dark, one fair, who seem to enjoy life thoroughly; but Clodagh did not enjoy Bob. What she particularly disliked about him—and wanted me to fix at once—was his habit of biting people in the rump while they were bent over cleaning out his hooves.

Like my Welsh ponies, Connemaras are quite big, about 13 hands high. Being a small woman, I can ride one comfortably, and they are perfect for older children and young teens. Bob was said to be very nice to ride and quite manageable under saddle; thus he should have made a fine addition to the riding school's mounts for children. The trouble he gave on the ground, however, was making them think twice.

Clodagh went into the stall with a halter and her grooming tools, while I leaned over the door and had a look at the pony. He was dazzlingly pretty: a dark bay (brown) with a black mane, tail, and legs. His flanks were dappled with golden circles or spots, like a child's painted rocking horse. He had big eyes, small ears, a beautiful lush mane and tail, dainty little feet, and a body all curves and graces, like an Arabian horse.

He was also, I thought, wearing a rather pleasant expression, but as soon as Clodagh tried to brush him he danced around, back and forth, left and right, swinging his teeth warningly in her direction. She put the halter on and tied him up short to a ring in the wall. He swiveled from side to side and fought the lead line, trying to get a nip at her as she picked out his hooves. When she finished with one side and went to the other, the horse swung himself next to the wall. Clodagh moved into the gap between horse and wall and told him "Get over," poking him in the ribs. It made me uneasy; an aggressive horse in that situation can mash you against the wall, and it's a hard position to get out of. But Bob was not dangerous, just naughty. He did move away, heavily and reluctantly, after she'd shoved him a good deal, and Clodagh threw me a "See what we have to put up with?" look. Meanwhile Bad Bob's pretty little feet kept up a samba, and sometimes he pawed at the floor, with great shoulder-high impatient sweeps of a front hoof that somehow *just* managed to miss hitting Clodagh's leg or foot on the way by.

Furthermore, when Clodagh turned away from him for a few seconds, Bob began biting at the lead line that tied him to the wall, studying the knot with eyes and ears and plucking at different parts of it with his mouth. "Look, he's trying to untie himself!" I exclaimed. One of my Welsh pony stallions had discovered how to do that. Clodagh sniffed disapprovingly and I agreed with her implicit opinion that the horse that can untie knots is a big nuisance…but smart!

So the training began. Jill had assembled quite an audience on my side of the stall door: some foundation staff, a few guests and board members, a television cameraman from Dublin, and a volunteer crew of about a dozen or more girls, ages 10 to 14, who work in the stables in a summer program, each neatly dressed in riding pants and navy blue Festina Lente shirts.

I decided to start with Alexandra Kurland's excellent first step, teaching a horse to touch a target, in this case a little blue traffic cone someone found for me in the stables. You can do this with the horse in its stall and you outside, so if he gets pushy or excited you can safely step out of reach. Meanwhile the horse learns what the clicker means, how to make you click, and how to control his own behavior instead of just reacting to yours.

I had a pocketful of alfalfa pellets, a favorite horse treat, and, of course, a clicker. I asked Clodagh to come out of the stall. Bob's head and neck came over the stall door after her. He got a click and a treat from me, saw where the treat came from, and instantly began trying to get into my coat pockets. I backed away a little, and held out the target. Sniff, click, treat.

In three clicks he was aware of the target and touching it on purpose. In ten clicks Bob could reach for it, move toward it, and bump any part of it with his nose; he was now target-trained. In fact he liked his target so much that if I tucked it under my arm or behind my back he began mugging me—reaching for me with his mouth and bumping me with his nose—trying to get at the target and get clicked.

The next step therefore would be to teach him to keep his nose and teeth to himself, out of my pockets and away from me. I'd started with a Kurland method of feeding him at arm's length so he had to back up and move away from me to get his treat; but I didn't think he had made the connection yet. That's Pavlovian, or classical conditioning: you associate two stimuli, in this case, the food and the location "away from me" by repeatedly pairing the two. But classical conditioning is slow, and takes many repetitions. The heck with that; I wanted Bob to mind his manners *now*. So I started clicking him whenever he happened to swing his head away from me. In three or four clicks he was deliberately swinging his head away, hearing a click, and swinging it right back to get his treat.

Now it seemed to me he was taking his treat (two or three tiny pellets) nicely, instead of grabbing it. Maybe he wasn't going to mug me again; maybe I didn't have to back up any more. I tested him by standing quite still. He nibbled at my treat pocket flap, but very gently, and then stopped. You could almost hear him think, "Oh, right, I remember," and he swung his head away from my body. Yup, you got that right, pony; click, treat.

To get him to keep his head away for a while, not just for one nanosecond, I started offering the target when he was standing with his head away from me. Now, instead of swinging his head away and then right back at me he had to take his head away and then wait for the target. That worked like a charm. "Hold still, don't mug the American lady, and she'll give me the target. Bump the target and she's bound to click and treat." Of course by delaying holding out the target, a second more each time, I could teach him to wait longer and longer.

Now we had "Keep your head away" as an operant behavior, one the horse was doing consciously and deliberately. Furthermore, I'd shaped the behavior I was after by reinforcing it with the target (a cue) instead of just the click and treat. Strangely enough, that makes the new behavior more reliable more quickly. It's counter-intuitive but very powerful; therefore a nice lesson for the audience, too.

This took perhaps 15 minutes. The whole time, Bob's pretty feet were moving impatiently under him; he was learning a lot and working hard, but he was also keeping up his samba. Now, suddenly, he lowered his head a little and stood still, with his ears, for the first time, relaxed.

"Oh now that's a novel idea," I said to him sarcastically, "Standing still," and I clicked and treated him for that. It was the first time I had spoken to him, and it wasn't for his benefit, just for mine. But I knew it was time to stop. He was tired and so was I, or I would not have been sarcastic.

I told the camera we were done for now. The kids were full of questions and I was happy to talk, but first I shooed everyone out into the stable yard to give poor Bob a well-earned rest.

The new mare

One of the Festina Lente coaches had just bought himself a new competition prospect, a three-year-old mare who, he told me, presented another problem: she was hard to catch and halter. Would I consent to work with her? Sure, why not. I leaned on her stall door and watched with interest as Helen Keogh, one of the staff, went in to halter her. The mare pulled away and lifted her head toward the rafters, but Helen deftly slipped the halter on anyway. Jim, one of the TAGteaching riding coaches, stepped into the stall and quickly picked out the mare's hooves. (At Festina Lente every horse must have its feet cleaned before leaving its stall. Thus not a speck of manure sullies the stable yard or offends the public eye. If some horse does commit a nuisance in the yard or the arenas, one of the uniformed girls is instantly there with broom and pan to clean it up.)

Helen led the mare out into the stable yard. The mare was a beautiful hunter/jumper type, dark brown, and very tall; I couldn't see over her back. The children and staff started to gather around, but much too close, I thought. I moved them all well away to the side. In fact I made sure that the whole stable yard in front of the mare was empty. She was new and she was nervous. If horses panic and bolt they go straight. Now if she lost her courage at least no one would get run over.

While Helen held the mare's lead rope I introduced her to the clicker. She did not seem to know how to eat food from a person's hand but she figured it out. I made sure that the first couple of clicks were absolutely simultaneous with the food-to-mouth so she quickly acquired that association. Now I tried another Kurland tactic: Question the horse. May I touch you on the shoulder? Yes? Click/treat. May I touch you on the leg? Yes, click/treat. May I touch you on the side of the neck? Yes? Click, treat. May I touch you on the neck near your ears? No, please don't, she answered, pulling up and backwards away from my hand.

I tried that again a couple of times, moving my hand lower on her neck, click, treat, and then back up toward her ears. Oops, no click, no treat. Not much progress. Hmm. For speedier progress, what you need is for the animal to be doing something you can react to positively, *not* for you to be doing something the animal has to react to.

I moved on to another Kurland plan: get this mare to lower her head, which might put her head within reach and which also seems to make horses feel calmer. One recipe is to lay your hand on the horse's poll, the flat place right behind the ears, and keep it there until the horse moves its head downward; then click and treat. Well, this mare was too flinchy for that, and besides, to reach her poll, I'd need a ladder.

What about just free-shaping head moves downward? Clicking little random movements?

As I thought about that, the mare happened to lower her head a little and I clicked during the move. Treat. Wait. Again she dropped her head a few inches, and I clicked it. Now she dipped her head on purpose. Three clicks later her nose was on the ground and she was snuffling around for treat pellets she might have dropped by accident. Hey, that was easy! I thought it was a good place to stop. Quit when you've made progress and had some success. Helen turned the mare around and took her back to her big, comfortable stall.

Jill Carey came over to me. "I'm glad you did something with the new mare," she said. "I feel so sorry for her, she's bewildered and away from home for the first time, and the pony in the next stall is making faces and being terribly rude to her. Yesterday she just hid in the back of the stall and never even looked out her door."

"Well, she's looking out her door now," I said. Indeed, she was hanging her head over the stall door, looking all around, studying the stable yard and the other horses with interest. Furthermore, she was lolling her long tongue out of her mouth and pulling it back in again in a most inelegant way. I laughed. "She's remembering about the food," I said, "and about earning all those treats! This stable yard looks like a pretty nice place, all of a sudden!"

But had she learned anything else? A bit later I walked over to the stall door. She didn't back away. I laid my hand firmly on her neck, up near the ears, just where I'd been trying to go before. She started to pull away, hesitated, and then moved forward and deliberately put her neck against my hand. Click!

Might not look like much to a bystander. But I was thrilled. She could already use her new understanding to overcome her fear, and she had just given me a behavior that had not actually been clicked. Way to go, mare!

Bad Bob revisited

After lunch, at Jill's request, I spent an hour or so playing the Training Game with all the little girls; they were clever at it, too. Then Jill suggested we work with Bad Bob again, but in the outdoor arena, in the sunshine. Good, he might like that; and I trusted him a bit more now, too. He was more likely to want to earn clicks than to want to take off for the next county.

I decided that the behavior for this session would be "stand still while you are being brushed." It wasn't clear to me why Bob fidgeted so during grooming and handling. Was it really uncomfortable for him? Seemed unlikely, since most horses love being groomed; in fact when we clicker train foals that are too young to eat solid food we use scratching or brushing as the reward.

Maybe the dancing around was one of those accidentally owner-reinforced behaviors like so many bad behaviors of pet animals. In fact, it didn't matter how it began. We could at least start to get rid of it. Jim Mernin brought Bob from the stall into the arena and held him by the lead line. Clodagh would brush him, and I would do the clicks and treats. The children gathered along the fence to watch.

I started by clicking for standing still during one stroke of the brush, and then two. Bob reached for me with his mouth once or twice in the beginning, and I thought I would have to go back and review "Keep your head away," but after a few clicks he settled down and just waited politely for each treat to be handed to him.

I asked Clodagh to stop brushing when she heard the click, and that helped a lot. Now Bob had two kinds of rewards for the click: a food treat arrived, and the brushing went away. Soon he could stand still for five or six brush strokes on his right side, and Clodagh was expanding the area she brushed, from his neck to his shoulder to his back and his rump. It was hard for him to just stand there. A couple of times one foot moved just as I clicked. Once, when the brushing stopped, he twitched his skin all over, the way horses do when a fly lands on their backs. "Tickles, does it?" I said, wondering if that was really so.

I moved Clodagh and her brush to Bob's left side and the war dance started up again. We had to go back to two strokes just on the neck, and review the training we'd done on the other side; but it did go faster the second time. Once during the session I asked Jim to give the pony a mental break by walking him around a little. Perhaps that helped; it certainly didn't hurt. Soon we were up to eight brush strokes on either side and then more. "Thirteen!" Clodagh exclaimed. That seemed like a lot, to me, so on the next try I clicked after just two strokes of the brush.

One of the beauties of clicker training is that it sometimes gives you remarkable glimpses of what's going on in the animal's mind. Bob stared at me with an expression I had never seen before on a horse: utter surprise. His ears were up, he was looking me straight in the eyes, and his face clearly said, "What! Just *two*? I can't believe you did that!"

"He's surprised, did you see that?" I said, laughing, to the watching people. It's easy to startle a horse; but this was not alarm. This was just pure amazement. That expression told me two things. First, he thoroughly understood the game we were playing. Second, in his past experience, things usually just went on and on getting worse, not better. What I wanted to do was to pat Bad Bob (at present he would probably hate that), or throw my arms around his neck, or give him a month's supply of alfalfa pellets. What I did was smile and pay him his treat. Poor thing. Maybe Festina Lente will turn out to be a better place for you.

And let's begin, I suggested, by changing his name. Why add to the bad karma he already has? So he went back to his stall temporarily called Bonnie Bob. Maybe he will become Robin or Bobbin or something else suitable for a nice friendly pony; I'm pretty sure there is one in there somewhere. He has become Clodagh's special project now.

The children

Jill had one more trick up her sleeve. She began with a question: did I think the girls could try some clicker training with the horses? Of course. Three new horses were immediately led into the arena: a regular riding school horse, a beautiful gray Irish hunter I coveted on sight (I have a weakness for grays), and Rosie. Rosie was something totally new to me. She was what's called a cob; easy to ride but sturdy enough to pull a cart. She was a filly, just two or three years old, a pinto, white and brown, with a naturally arched neck, a pretty face, and beautiful white "feathers" on her legs—fetlock hair that started almost at her knees and fell so thickly you could hardly see her hooves. Jill told me that she had spotted Rosie tethered on a bit of wasteland on the other side of the high garden wall and bought her from some gypsies, or travelers as they say in Ireland. And the day after she bought Rosie, on the other side of the garden wall, there were six more just like her.

While the horses were coming in, I found three targets: Helen had one she'd made herself, I had a little folding target stick in my purse, and the girls could have the blue traffic cone they'd already watched me use. I divided them up into three groups, six or more children in each group. While adult staffers held the horses' heads, each group would teach their horse to touch the target. I invented a procedure to give everyone a turn: "One girl offers the target and does the clicking; another girl treats. You each get five clicks, and then the trainer becomes the feeder and someone else gets five clicks. When everyone has been both the trainer and the treater, we'll see how you're doing."

Hey, just like my Welsh pony days! I tell the kids what the day's assignment is, and the kids train the ponies. Easy for me! I relaxed and chatted with the onlookers, and learned that Helen was still marveling at how long Bad Bob had been able to stand still.

When I saw that everyone in the three groups seemed to have had at least one turn as trainer, I went round and gave a test. The horses were touching the targets, but would they follow a moving target? Sure enough, as I held the target out and started forward each horse pointed its nose at its target and came along with me. Brilliant! Great job, girls.

I thought that was the end of it, but Jill let them go on. Now each group of girls collectively thought up a new behavior and trained it. The grey hunter learned to walk backwards when the target was placed behind his chin. The school horse learned to circle, if I recall correctly. And Rosie was taught, with the target, to stick her nose both down to the ground and straight up in the air as far as it could go.

We had a debriefing, with the girls all in a line. There seemed to be more of them than ever, now there were 18 at least. "What do you want to train the horses to do next?" I asked. They thought about that. Then Megan said, "Stick their tongues out." Yes, all the girls seemed to think that would be feasible. Fine with me, I said. (I must say, later on when I envisioned a whole stable of horses looking over their stall doors with their tongues hanging out, I hastily e-mailed Jill. If the girls did that, they should be sure to teach the horses to put their tongues back in again, too.)

The outcome

So what was the score?

I had a blissfully happy day of "playing ponies."

Bad Bob had learned that sometimes it pays to try to be good. Let's hope it leads to a total reform.

The new mare had learned to be less afraid of her new world and to take on some responsibility for her own well-being.

Three other horses learned to follow a target and undoubtedly are hoping to play the clicker game again.

And there are now approximately 18 new young clicker trainers in Ireland.

The Uses of Pandemonium—A Trainer's Take on *Good Morning America*

> Like the scared mare on the previous pages, a naïve puppy
> learns that she can not only cope but learn to thrive during the chaos
> of a live TV production. The clicks are clear communication in an
> incredibly noisy, busy, crowded environment.

I've been invited to appear on *Good Morning America*, to promote my new book, *Reaching the Animal Mind.* I've offered to give a demonstration, if they can find a suitable dog from a shelter.

Sandra, an executive with the New York Humane Society, sometimes takes adoptable dogs on *Good Morning America*. Sandra tells me it's usually just she, the dog, the hosts, the set, and the cameras.

But this time it's not like that.

The shelter is providing exactly what I requested for the GMA segment—a bouncy, friendly, young dog with little or no training. Mia is a 5- or 6-month-old purebred golden retriever. She was turned into the shelter by a family that had to move. Mia was adopted just two days ago by a pretty, young veterinarian named Nicole. Nicole will lend her back to us for the show.

The show is scheduled for Tuesday, June 16, 2009. I have agreed to meet Mia in a shelter exercise pen on Monday. GMA has sent a camera crew.

I ascertain that Mia likes hotdogs. She knows her name. She's sweet and gentle, but lively enough to make a good demo dog. The camera rolls while

Mia learns about the clicker, gets the idea of hand targeting, and starts offering a sit.

After 10 successful minutes (nearly 80 clicks and treats), she is weary and goes to the gate, looking for her new owner. "You bet, dog. Go home."

"We're done here," I say.

"No, no," says Joe the camera guy. "I'm not through; I need more shots."

I refuse. "This is not a performing dog. This is a puppy. She's tired."

Puppies are great at clicker work, but they get tired very, very quickly. Joe, however, is desperate for another view of the sit. I give in, reluctantly, and let him film two more sits over my shoulder.

At the ABC TV studio

The next morning the GMA limo delivers me to the ABC-TV stage door at 5:30 a.m. The plan is for the dog, her owner Nicole, and Sandra from the shelter to wait in a separate, and quiet, room until the 6:30 a.m. rehearsal, and then again until the 8:30 a.m. show. I don't want the puppy to see me until we're actually training. I don't want her to glimpse me and get all excited and worn out.

It's a long trip from the waiting rooms to the set. The dog and her handlers go in the elevator. To stay out of sight of the dog, the producer takes me up and down the stairs. We will all make this trip not twice, as I expected, but six times!

The first trip turns out to be a false alarm. The second trip is for the rehearsal. The set consists of a scrap of carpet, a couch, and a table for my hotdog container, all at one end of a big TV studio with windows onto the street.

We rehearse, with the producer standing in for the host. The dog comes on set and tries to get into the producer's lap. A hand target gets her off—good. Mia jumps on me a few times. She locates and dives for the treat bowl, but she listens to her clicks and, by and by, comes, sits, and hand targets nicely. It's a good training session, I think.

Uh-oh. Then, and I did *not* expect this, we have to go *three* more times for what's called a "bumper," a short, live scene of me and the dog on the set. We do this so that Diane Sawyer and Robin Roberts, outside somewhere, can talk about what's coming next, namely us.

The dog, of course, is getting tired. On about the fourth trip downstairs, while I'm keeping her busy until our five-second bumper is over, Mia loses focus entirely. She doesn't react to the click, she skips eating the hotdogs, and she's beginning to roam. Uh-oh.

I detect one source of the problem. It's not just fatigue; it's *way* too much stimulation. As we all head back up two flights, there are *many* more people around now—and they want to pet the dog. As the freight elevator door opens, revealing Nicole, Sandra, and Mia, I see a large man who is actually on his knees, rubbing and mauling the puppy's head and ears, proving what a big dog lover he is.

The elevator group heads past us down the hall. Mia lies down in a doorway and another man lies on the carpet nose to nose with her, mugging the patient but bewildered puppy. I actually run down the hall to catch up. I tell Nicole and Sandra not to let people pet the puppy anymore, because it's wearing her out.

I think they must have succeeded. For the next bumper session Mia is aware of her clicks and focusing again.

In the sound studio

Meanwhile, on each trip downstairs the stress and distractions get worse. I expected the huge cameras, lumbering around the studio like robotic dinosaurs with their handlers and their black, ropy innards trailing behind them. I did not expect the rock band. Ashley Tisdale from *High School Musical* is performing this morning. The stage occupies the other half of the studio. Between bouts of actual music, the band is emitting occasional warm-up blasts of incredibly loud sound. The music has the camera crews dancing—and me, too. It's very infectious. But what is all this noise doing to the poor dog?

The next time a huge noise erupts, I click. I see Mia's head snap round to me; she heard that! I run over to the dog and give her a treat. I see her relax a little.

"Ah ha! So, hearing a Very Loud Sound is a clickable moment? Well then." Pretty good for a puppy!

I was told there would be no audience, but on trips four and five the public has been let in. People are standing four and five deep all around the room, right up to, and even on, the little bit of carpet where the dog and I will work. Some are in costume. Some have signs, some have small children, some are sitting on the floor, and some are carrying food. It's like being in the middle of a parade, or backstage at the opera: jam-crammed chaos. As the final touch of pandemonium, a bomb-sniffing dog is threading through the crowd's legs, working the room.

The music starts up. Ashley belts out a song. The boom camera, right in front of me, sails its long neck and head through the air around the singer, near and far, above her, beside her—my goodness, what that guy can do with this machine! Who gets to see this? I stop worrying. This is enormous fun.

Showtime

Trip six. It's performance time. Nicole and the puppy have been stationed right next to the carpet, to make a quick entrance. Mia can handle this now.

"Sometimes I'm with the clicker lady; sometimes I'm with my person." She is calm.

The host, Chris Cuomo, has arrived. Naturally, he plays with the clicker. I dash over and hand Nicole a couple of treats.

"If he clicks, you pay Mia, okay?" With each of Chris Cuomo's random clicks, Mia gets a treat.

Chris chats with me off camera, mostly telling me how he trains his Rottweilers. I have no comment. The viewing audience sees video of me coaching a condor trainer in a zoo. Then it's our turn. The host makes some jokes, aiming the clicker like a remote. He then describes the assignment I had, "training a wild, uncontrollable puppy in one day."

"No, in ten minutes," I say. Oh.

Meanwhile, on the words "wild uncontrollable puppy" the camera switches to Mia. She is flat on her belly, chin on the floor, nose pressed against her person's foot, falling asleep.

But she wakes to my voice, comes onto the set, and performs perfectly for two full minutes, providing a wonderful demo, and giving me lots to talk about. At the end, I suggest that the host click the dog. He sticks out a tentative hand. Mia promptly bumps it. He clicks and I treat. "I know this," says the dog. "I get this," thinks the host.

He leaves the set clutching a signed copy of *Reaching the Animal Mind* because he wants to read the part about parenting. Good!

It's a convincing demonstration, friends tell me later. If they only knew what was going on all around us, and what the puppy had already been through before trip six!

One last click

I pack up and go downstairs again. The producer leads me toward the stage door through yet another mob of people. A football sails over my head and I turn around to look. Our host, behind me, is passing the ball back and forth with someone in another part of the crowd.

The ball flies over my head back to him again. Just as he reaches up and catches it, I click. He startles, looks at me, and calls out, "I wish I'd had that in high school. I'd have caught a lot more passes." Yes! He's got it!

Clicking works; I know that. But once in a while it's almost a shock. I felt it on the set with Mia, and now I feel it here. Oh my goodness. Look at that. It worked again!

What Do Dolphins Do for Christmas?

In contrast to naïve animals, clicker-savvy dolphins
all but train themselves.

Christmas in Hawaii is always a little different. Santa might wear flip-flops instead of boots, a red *pareu* (sarong) around his *opu* (stomach), a red hat, a red lei—and nothing else. At Hawaii's Sea Life Park, where I was head trainer for a decade in the '60s, we put on dolphin shows many times a day and sometimes had big crowds of school children. Naturally, we thought of Christmas-type events: dolphins pulling Santa's sleigh—with gift-wrapped buckets of fish and a Hawaiian *poi* dog riding on top of the sleigh—that kind of thing.

My favorite Christmas show took place in the all-glass Ocean Science Theater, where you could see the whole tank under water. One year we decided to teach Hou and Malia, two rough-toothed dolphins, to trim an underwater Christmas tree. We put the tree in a weighted pot in the center of the tank, on the bottom. We gave each dolphin a long, gold garland. Holding one end of her garland in her mouth, Malia, the senior dolphin, swam in circles around the tree, from right to left and bottom to top, leaving the garland neatly wound in a spiral up the tree. Then we gave Hou another garland, and she circled the tree from left to right and top to bottom, leaving that garland neatly crisscrossing the first one. Then, if I recall correctly, the dolphins took a bunch of plastic ornaments and put them on the tree, too.

Rough-toothed dolphins

Rough-toothed dolphins, *Steno bredanensis,* are homely, blotchy brown animals with ugly faces, not at all cute like Flipper, but they are incredibly smart and very interested in learning. Trainers used to joke that all you had to do was to write the show plan on a blackboard and hang it in the tank. (SeaWorld trainers have the same joke about killer whales.)

Here's a sort of summary training session:

Dolphin: "You want me to do what? Go around that prickly thing [the tree] down there?"

Trainer clicks (only we used a whistle, which carries better under water).

Dolphin (swallowing fish and ready to try again): "Which way shall I go around? This way?"

Click.

Dolphin: "And starting where?"

Trainer gives a "go" cue as the animal approaches the bottom.

Dolphin: "Okay, got that."

Click/treat. Repeat.

"And now you're giving me a thing to carry?"

Trainer hands over the garland, animal takes it. Click. Fish.

Now a little time is allowed for the animal to get used to swimming around dragging the garland—oh, say, 10 seconds. You don't want the animal to get too interested and start playing 101 Things to Do with a Garland.

Finally, give the garland, give the "go" cue, animal wraps it around the tree. Train animal number two. Put it in the show. Audience loves it. Merry Christmas!

The hardest part was getting the tree rigged so we could lower it from the ceiling into the tank when it was needed, and then raise it up out of the way again. That required getting Maintenance to show up with a crane and fasten ropes and a pulley to the Ocean Science Theater ceiling. Getting that 20-minute job onto their schedule took *weeks* of struggle.

The Panda Game

Miniature guide horse Panda can not only train
with the best of them but can express herself in
ways individual and sophisticated.

Miniature horses are a special breed. According to horse owners, miniature horses are not descended from ponies, but developed from regular horses. Most of them are about the size of a large dog, and, like some large dogs, they make great guides for blind people.

How can that be? Well, first, yes, they can be housebroken. Like a dog, a horse can learn where and when to relieve itself, and how to signal when it needs to go out. A little horse can also easily learn to carry out all the functions of a guide dog: watching traffic, stopping and alerting the owner at curbs, steps, and stairs, avoiding obstacles, staying quietly next to the owner in restaurants and other public spaces, and so on. A guide horse can ride in cars and on trains (not planes, or at least not yet). A guide horse and its owner quickly become personal friends, and learn not just to work together but to play and relax together at the end of the day.

And here's the money part. Miniature horses live a lot longer than dogs. Once trained, a guide dog has six or seven years left before it is too old to work. The owner of a guide horse can expect to enjoy the services of this faithful friend for 20 years or more.

There are, I'm told, about half a dozen miniature horses functioning as guide animals in the US right now. One of them, a black-and-white mare named Panda, belongs to Ann Edie, a teacher in New York State. Panda's trainer was Alexandra Kurland, and Panda is the first guide horse to have been 100% clicker trained.

KPCT was fortunate to have Ann Edie and Panda as honored guests at ClickerExpo Newport in 2006. Everyone enjoyed meeting this distinguished pair. We were awed by Panda's calmness as she guided Ann during the day, through crowds and halls and past all sorts of dogs (some of which were distinctly upset at having a horse among them). People were wonderful about not trying to pet Panda as she worked, even though she is deliciously cute and furry. At the Saturday night autograph party Panda even signed her own books, *Panda: A Guide Horse for Ann,* with a little inky front hoof.

During plenary sessions, the nearly 400 attendees had a chance to hear Ann's eloquent comments on the relationship between them, on Panda's skills and intelligence, and on the many ways Panda shows Ann that she is happy—from her greeting whicker in the morning to the ways she likes to play and relax in the evenings.

The morning after ClickerExpo, the faculty gathered for a session of brainstorming on next year's ClickerExpo, and of sharing our personal news and events. And what Alexandra and Ann shared was a participation exercise: something that Panda had invented for herself called the Panda Game.

This is how you play the Panda Game. All the people present were given a handful of Panda treats: bean-sized alfalfa pellets, much enjoyed by horses. About a dozen of us joined in a big circle, standing about an arms' length apart.

"What do we do now?" we asked.

"Panda will show you," Alexandra said. Alex removed Panda's guide harness and told her to start playing.

In a businesslike way, Panda set off for the first person on her left—Emma Parsons, circled behind her, came up alongside in "heel" position, and halted. Emma's response was instant: Click, treat. We all learned to hold the treat under Panda's mouth, since Panda is trained to wait for her treat to be fed to her. We also had to learn to hold our hand flat, so she could nibble up the pellets without nibbling one's hand by accident.

Panda then briskly moved to the next person, circled behind her, came up to her side, and again earned a click and a treat. The third person was Kay Laurence. This time Panda came in and halted at a 45-degree angle, instead of straight. Kay instantly stepped sidewise, conveying the information, "None of that carelessness from you, my girl," and Panda adjusted her position just as quickly, and got a click and a treat.

Panda went on around the circle, methodically teaching each of us to click and treat. Then she started around the circle a second time. This time, being the primates that we are, always curious and restless, people began introducing slight variations. When Panda came along side Ken Ramirez, training director of the Shedd Aquarium, Ken took a step forward and stopped. She matched him exactly with her own steps, and got a click and a treat. When she got to me, three people later, I tried two steps forward. Panda came with me with precision, heeling like a high-scoring obedience dog.

Then she got to Aaron Clayton, our company president and Clicker-Expo host. Ever the risk-taker, Aaron took six steps forward. "No!" said Panda. She tossed her head, switched her tail, and trotted away from him across the circle, to join Kay Laurence instead. We all burst into laughter, of course.

Now the game became more interesting. Instead of going around the circle taking each person in order, Panda began crisscrossing the middle, choosing whom she would play with each time. Those who were favored by being chosen began asking for more behavior—back up a step, say—but tactfully, since we had now witnessed Panda's ability to express her opinion.

Panda seemed to prefer the skilled, but she was not a snob; she also rejoined those people who were still a little awkward at feeding the pellets. After a few more stops she gave Aaron another chance. He took her a modest two steps forward and clicked. Panda accepted her treat and moved on. Did she look a little smug? I thought so.

By and by, the game began winding down, as some of us used up our supply of treats. A few people were still playing when Ken Ramirez and I happened to start walking back to the meeting table together—until Panda barged between us, circled behind Ken, and actually herded him back toward the game.

Ken, for those who don't know, is a leader in the marine mammal and zoo training worlds and a hugely innovative and advanced trainer. We all think he's The Best. Apparently Panda thought so, too. Ken, being gently jostled by this determined little creature, looked at me with awe on his face and said "I feel so flattered!"

"Indeed, you should!" I exclaimed. Alex gave Ken a few more pellets so he could play another round or two of the Panda Game.

One of the joys of having clicker trained animals around is that they are able to express themselves, not just with innate social behavior, the way animals usually do with people—infantile demands such as whining or begging, species-specific social displays such as threats (growling for dogs, laying the ears back with horses)—but with their whole intelligent selves. It's interaction on a new plane, one that those who dominate, punish, and suppress behavior, and call it training, will never see.

And Panda just gave us a world-class example. In a game she invented herself. A game that earned her not just our affection, but something rarer, both personally and intellectually—our respect.

The Art and Science of the Clicker

The science is clear: click means "Bingo!" and winning is always a high. Neuroscientists now have proof of the primitive, emotional power of the click. While the principles and mechanics of clicker training are straight-forward, Karen shows us that practicing them takes as much art as science. If you're not going to push, prod, or lure an animal into doing what you want but instead are going to build on whatever the animal is offering you, you have to use your imagination and relearn to see: What can your animal do, and how can you mesh that behavior with what you want it to learn? Can you train your brain to recognize the little pieces of behavior that go into shaping? Karen walks you through some exercises to sharpen your eyes and mind.

"Look, ma! No hands!" Terry Golson shaped her horse Tonka to "heel"—no lead rope necessary (see page 79). Photo: Steve Golson

Clicker Training vs.
Lure-and-Reward Training

Most of us have heard the criticism that luring a dog to do a
behavior slows learning because it blots out thinking: the dog
simply focuses on where the food leads him and doesn't register
what he's doing. But did you ever think whether luring might
eclipse the clicker as well, or what sort of dog luring produces?

Amidst all those Association of Pet Dog Trainers (APDT) teachers and instructors at their convention in 2004, I thought a lot about teaching: teaching the art and science of the clicker. ClickerExpo faculty member Kathy Sdao has given this topic a lot of thought, too. At APDT she presented a socko, four-hour workshop titled, with a tip of the hat to poet Ogden Nash, "Candy is Dandy but Clicker is Quicker," on the *real* differences between clicker training and lure-and-reward training.

Why is clicking a more powerful way to teach a behavior than walking the animal through the moves by holding a piece of food against its nose? In clicker training, the animal is thinking about what behavior caused the click, and how it can do that behavior again so it can get some food. In luring, the animal is thinking about the food, and following the food until it gets some. Whatever else it might happen to be doing—sitting, lying, heeling, turning, performing obstacles—is overshadowed by the presence of the food. If you happen to add a clicker, that's overshadowed, too, Kathy told us. The food has been there before, during, and after, so the clicker makes no particular difference. It doesn't tell the dog which behavioral moment actually earned the eventual delivery of the food.

I saw quite a few lure-and-reward-trained dogs at the three 2003 ClickerExpos, and I soon noticed that in some ways they were all alike. They were docile. They sat or lay quietly when lured into position. They walked quietly wherever their owners went. They usually didn't lunge at strangers or try to play with other dogs. They were compliant; but they were also rather unresponsive. Owners were apt to steer them with the leash or lure them with food rather than speaking to them. Often the dogs seemed not to recognize verbal commands.

It wasn't a bad arrangement. The dogs got food now and then all day long, and the owner got a compliant, rather inactive dog. I can see why lure-and-reward training has become popular. Compared to the most widely used alternative—traditional, correction-based obedience-type training—it has many advantages. It's easier to teach. It's easier for the pet owner to learn. You don't need so many physical skills. You don't need to "dominate" your dog. And it's *much* easier on the dogs.

But. It's not clicker training. The dogs aren't "operant"—trying to learn, able to understand and communicate, offering behavior (including sitting, lying quietly, walking at one's side) with confidence and understanding. And that's not the fault of the students but of us teachers.

Click vs. Voice

Luring doesn't tell an animal what it's doing right.
When animals know precisely what to do to get what
they want, they learn faster—a lot faster.

The click is clear. The click means one thing only: "Bingo. You win."

It's what you feel when you're waiting for a special call and you hear the phone ring. It's not the call itself; it's the clear-cut, simple message: "You got it."

The click is more than just a conditioned reinforcer. It's what engineers call "feedback." It marks or identifies precisely which behavior is paying off and should be repeated. That's why we call it a marker. (FYI, this use of the word "marker" is not a clicker training invention. I first heard the phrase "event marker" from Skinner's protégé, the late Ogden Lindsley, who used a sound as a marker signal to teach his pet donkey to open a mailbox. Ogden probably learned it from earlier behavior analysts.)

The voice is a wonderful tool for conveying the trainer's pleasure and pride, and for giving warnings, instructions, and cues—and we use it for all those purposes. But because it has all these functions, the voice does not make a very good marker. The brain must filter out all that other information.

Because the click marks or identifies precisely which behavior pays and therefore is worth repeating, it seems to be that "clicker is quicker." We know this from practical experience. For example, the complaint I heard many times in the early days of clicker training's growth in the dog world was this:

"I converted all my classes to clicker training, and now my students are going through my six-week curriculum in four weeks." That gave the teacher a new problem—how to keep the students busy during the extra two weeks they'd paid for!

But is this difference in speed of learning real? And, if so, is it due just to the click? Psychologist and clicker trainer Lindsay Wood decided to take a look at the question experimentally. She trained a number of naïve shelter dogs to cross the room and bump a target, using the click as a marker, and taught another group of naïve dogs to perform the same behavior using the word "Yes" as a marker. Same trainer, same skills, similar clueless dogs, but the clicker dogs learned the behavior roughly 40% faster.

And Lindsay teased out another very interesting piece of information—the clicker only made that difference during the learning phase. The word "yes" was as good as a click for maintaining the behavior once the dog already knew what to do. It was only when the dog was trying to figure out what worked that the click displayed its power.

The next big question, of course, is *why* the difference occurs. That's due to how reinforcers are processed in the brain. Neuroscientists know something about that.

The Amygdala: The Neurophysiology of Clicker Training

German scientist Barbara Schoening is a clicker trainer and a veterinary neurophysiologist in private practice. It was she who first drew Karen's attention to the relationship between clicker training and research on stimuli and the amygdala, a structure in the limbic system or oldest part of the brain.

Research in neurophysiology has identified that certain kinds of stimuli—bright lights, sudden sharp sounds—travel to the amygdala first, before reaching the cortex or thinking part of the brain. For instance, a loud crash behind us makes us jump, before we even think about it or turn to see what it was. It's our ancient, automatic alarm system at work. Once the click is turned into a conditioned reinforcer, it also becomes that kind of automatic stimulus. Conditioned fear responses in humans are established via the amygdala and are characterized by a pattern of very rapid learning—often on a single trial—and long-term retention, with a big surge of concomitant (negative) emotions. We clicker trainers see similar patterns of very rapid learning, long retention, and emotional surges—albeit positive ones rather than fear.

Barbara and I hypothesized that the click is a conditioned "joy" stimulus that is acquired and recognized through those same primitive pathways through the amygdala. This might help explain why it is so very different in its effect from a human word. The click is a conditioned reinforcer with no meaning other than good news. Any word a human says, however, has many previous associations, and not all of them positive. Is this "yes" for me? Is it good news or bad? Is the person mad at me? It takes an instant of attention and perhaps some repetition to be sure what the message means this time.

The consequences

If the click is processed by the central nervous system (CNS) much faster than any word (such as the word "yes!"), that might help explain why clicker trained animals acquire new behaviors so quickly. After all, even the most highly trained animal or verbal person first must recognize, and interpret, a word before it can "work." And the effect of the word may be confounded by the emotional signals, speaker identification clues, and so on, noted above.

We've already seen that clicker classes for pet owners typically cover the standard curriculum in much less time, with a higher degree of success than do traditional classes, so that instructors usually go on to tricks, introduce agility, or move into their intermediate curriculum to fill up the weeks students have paid for. Perhaps the dogs learn faster and more accurately, but the people also get feedback from the clicker. It increases their attentiveness to the dog, improves their timing, and for all we know, triggers nice feelings in the amygdala.

There are many additional possible neurological and biochemical side effects of clicking. Here's a comment from Pat Robards, clicker trainer and editor of *Dogtalk Magazine* in Australia:

> *Humans experience episodes in which the parasympathetic nervous system (PNS) is active as nice warm feelings, relaxation, contentment. Any time that a previously neutral stimulus, like a clicker or a kind word, gets paired with one of these parasympathetic reactions, through classical conditioning, the clicker acquires the ability to produce the same pleasant effects. This is why treats (and soon the clicker) can be used to calm a dog, make him less fearful, and cause the whole training process to be a happy experience. That's one of the reasons clicker training is at the cutting edge! I use it to mark Calming Signals for a fearful dog, thanks to Karen Pryor.*

The proof

So, yes, clicker is better. But there's more. As I reported in Chapter 10 of *Reaching the Animal Mind*, neuroscientists consider that *all* conditioned reinforcers, such as the click, in fact *do* go straight through the amygdala—bypassing the cortex, the thinking part of the brain—to sites for memory and emotion. So the rapid learning, long retention, and joy we see in the animals we clicker train is real—and so is the difference in the responses people get from teaching animals by verbal instruction, luring, coercion, and punishment.

Clicker trainers tend to have strong objections to punishment. Now, thanks to the work of scientists like Jaak Panksepp, a big fan of clicker training by the way, we have a scientific rationale for why mixing correction and reinforcement is harmful rather than helpful to learning. It's not just a moral issue; it's common sense. Correction or rebuke switches the learner from the hypothalamus and its SEEKING mode to the amygdala's path of avoidance and fear. Clicker training combines both the excitement of the quest and the "Eureka!" moment (the reinforcement of the click). Keeping the whole process intact produces the happiest, quickest learners.

Charging the Clicker

Do you need to "charge up" the clicker before you can start
training? Karen points out that, in efforts to eradicate any
superstitious behaviors* in their animals, scientists in labs have
engaged in some superstitious behaviors of their own.

At an annual meeting of the Association for Behavior Analysis, where more than 2,000 behavioral scientists gather each year, a woman professor with whom I was acquainted told me that she had organized among her students a Rat Olympics. I was excited! What a good way to interest students in operant conditioning!

Alas, when I saw her videos, I was disappointed. Students were luring rats to climb ropes by holding food in front of them. They were baiting or shooing rats through tunnels. They made rats jump by placing a rat on a wooden bridge with a gap in it, placing food on the other end of the bridge, and then gradually widening the gap.

"Why didn't you use clickers?" I asked the teacher.

"Oh, they didn't have time! They only had a few weeks," she responded. What? Time for what? All this luring and shooing was a pretty slow business, after all.

Time, she meant, to condition the clicker.

---

* A superstitious behavior is an irrelevant, superfluous response that gets attached to a behavior erroneously during shaping, such as when a dog bows, then touches a target when the trainer just wanted the targeting behavior.

The logic of science

Science is supposed to be logical and based on proven facts, but scientists are humans, after all. They develop customs and practices in laboratory work that are based on nothing more than history, opinions, assumptions, and even superstitions. This idea that developing a conditioned reinforcer is a complex and difficult task, and that it must precede any "training" you will do by means of a conditioned reinforcer, is an example—it's not a fact, it's just a laboratory tradition.

Clicker training, or Skinnerian shaping as we practice it today, involves two kinds of conditioning. The first is classical (Pavlovian) in which some association is made more or less unconsciously between a stimulus (a particular smell, say) and what it makes you think of (pizza, your boyfriend, new cars, the dentist's office). Pairing the click with a treat, therefore, initially simultaneously and then sometimes with an occasional delay, click-then-treat, creates this Pavlovian conditioning, plus some tolerance for a slightly delayed treat.

The second kind of conditioning is operant. If I, the learner, do a particular thing, I can make the click happen. I press the button on the parking garage machine, a ticket comes out and the barrier swings up, letting me into the garage. This is a conscious association; the learner deliberately engages in a repeat of whatever it did before, in expectation of a reinforcer. That's what Skinner meant by the word "operant." The learner is the operator; the learner runs the machine.

What the clicker community wants

In the case of our learners, we want them to offer whatever behavior we clicked, hoping to make a click happen again. We first create the classical conditioning association, by clicking and instantly delivering a treat two or three times. When we see that the animal has noticed the food and is eating it and looking for more, we immediately choose some specific behavior to click. It needs to be something the animal is already doing anyway: looking at the trainer, say, or maybe looking at or smelling a target object we've presented. We click and treat during that behavior, several times in rapid succession. Now, right away, we are adding in the operant conditioning, using the clicker to mark some particular behavior as it is happening.

At some point the learner figures it out and begins offering the behavior "on purpose." Now we have a properly conditioned reinforcer: in the Pavlovian connection, "Click means treat is coming." In the Skinnerian connection, the click means, "What *you* did, made me click." The click marks, or identifies, a new operant behavior. And we did it all in the very first training session, probably in the very first two or three minutes.

Furthermore, in that very first session, we don't just sit there reinforcing the newly learned behavior over and over again. God forbid the animal should think there is only one way of making clicks happen. So we might go on to reinforce other behaviors as well as to shape additional criteria for the one we started with.

For example, in quieting a noisy kennel at a shelter, I might organize a few volunteers to go up and down the line clicking and treating any dog with its mouth shut (a quiet dog). And then, while some dogs still need to be clicked only for not barking, for dogs that are now quiet, I might suggest that the volunteers click and treat any dog that is not jumping (any dog with four feet on the floor).

When most of the dogs are a) quiet and b) not jumping, I might move volunteers on to clicking for eye contact. Most of the dogs would now tend to be looking into people's faces anyway, as the volunteers pass up and down the line of cages: "See, I'm quiet, see, I'm standing still, click me, click me." Looking up to see a person's eyes tends to lead to sitting, and bingo, we can now capture sitting, too.

By the time a dog that was once barking and bouncing around is sitting quietly at its cage door, that dog has learned not just one but four different ways to make a person click. It has "learned to learn." It has also discovered that giving one's attention to humans can really pay off; and it is well on the way to being adoptable. (This procedure takes about 10 or 20 minutes, depending on how many dogs and volunteers you have, and this learning can be permanent, needing only to be refreshed briefly and sporadically, especially when new dogs come in.)

What lab scientists practice

What happened with rats and pigeons in the operant laboratory was quite different, I think. The investigators only needed one behavior, pressing the lever or pecking a key. The learners were often required to do it over and over, many times, for each food reward. The investigators were not, as a rule, interested in variations of behavior—quite the opposite. If a rat associated some unrelated act, backing up, say, with the delivery of food, that "superstitious" behavior would interfere with the lever pressing and might skew the experimental results.

To avoid this inconvenient accident, researchers made sure, with a minimum of 200 reinforcers delivered randomly, that any accidental associations of behavior and food delivery were deliberately deconditioned, or extinguished, by going unreinforced in a blizzard of deliberately random events.

So this became the rule, in laboratories and textbooks and college classes: you have to go through an elaborate conditioning procedure to develop a conditioned reinforcer before you use it. My dolphin training teacher, Ron Turner, a graduate student of behaviorism at Columbia University, told me that in 1963. You *must* make sure the dolphin gets a whistle and a fish at least 200 times, with no particular associations, not at the same time, not in the same spot, not in association with any other behavior, to avoid the development of superstitious behavior. What a pain in the neck, especially with a newly caught dolphin that wasn't eating very well anyway, and might not take more than 10 or 20 fish in a day.

Does it work?

And actually, it was a waste of time. You go to all this trouble to teach the animal that the stimulus you're using, the whistle, or the click, or the blink, or whatever, means *nothing* except "food is coming." And then you want to turn around and ask the animal to learn the very opposite: "No, hey, now the click means you get food if you do that again." Maybe it made sense in the laboratory. In our world, where we want a huge variety of operant behavior as soon as possible, and we want it all on cue (under stimulus control, which is another level of operant behavior) as soon as possible, we want the animal to know and value its own magnificent ability to Make People Do Stuff from the very beginning.

You do have to maintain that Pavlovian connection, as Jesús Rosales-Ruiz and his students have elegantly proven in their research. People who make a habit of delaying the treat, of not having treats handy, or sometimes forgetting to treat, of substituting petting when the animal hates to be petted, may find that the clicker becomes less meaningful because the automatic conditioning has deteriorated. Often the giveaway, with dogs at least, is that the animal fails to stop what it's doing when it hears the click. Instead, it continues the behavior, while watching your hand; the hand movement of delivering the treat has become conditioned instead.

So "charging up the clicker" is an integral part of the first training session; perhaps in later training sessions you might begin with two or three closely paired clicks and treats, just to keep the Pavlovian conditioning strong. And deliberately "charging up the clicker" by giving 10 or 20 clicks and treats in rapid succession, unrelated to any particular behavior, is a remedial technique one might use with a dog that has stopped responding to the clicker because of some accidental deconditioning by the trainer.

But any long, randomized process you may have been told about is, I think, an artifact—a useless leftover of a laboratory procedure that is due more to custom (where did the magic number 200 come from, anyway?) than to science, and that in any case undermines the operant learning, and the important function of the clicker as event marker, which is the whole point of clicker training.

It's not What You Do. It's How You Do It.

For successful clicker training, you need to start with a behavior
that's already happening, not with something entirely foreign.
After all, if the animal isn't doing the behavior, you don't have
anything to click!

It's a sunny spring afternoon in New England at the annual meeting of a group of animal behaviorists. I'm sitting on a folding chair alongside a small corral, watching some clicker training demonstrations with horses. My friend Tim Sullivan, curator of behavioral husbandry at the Brookfield Zoo in Chicago, is sitting on my left.

A calm, old police horse is led into the corral and turned loose. The trainer stands on our side of the fence. She has a clicker, a bucket of feed, and a huge target stick that looks like a toilet plunger, with a big padded lump on one end.

The horse comes over to her. Click, treat. The trainer picks up the target stick, swings it over the fence, and jabs it at the horse. The horse has seen a lot of things in his life, but he's never seen this. Startled, he throws up his head and backs away. I laugh.

Then I'm embarrassed. I don't mean to humiliate the trainer; I should have kept my laugh to myself. The trainer coaxes the horse over again, shoves the target stick at him again, and startles him again, but less. In a few more tries she's got him touching the target, and the fear, though still visible, is almost gone.

Tim Sullivan leans over to me and says, "You can always tell when a trainer is coming from method instead of principle." Boy, can you ever! Golden words.

Method vs. principle

This trainer knew the method: present the target, click for touching the target, then treat. Repeat.

But, she was not relying on principles. One principle of shaping is that it is necessary to begin with something the animal is already doing—in this case, approaching the trainer. Nosing the end of a suddenly appearing, large, unknown object was definitely not in the existing repertoire. One would shape that event from a much simpler starting point.

Here's another fundamental principle of training: fear decreases existing behavior and increases avoidance behavior. Did the trainer want to slow down the horse's learning of a new behavior? No. Did she want to develop avoidance behavior, such as tossing the head and backing away? Not really. She had a method, but not an understanding.

Watching, my take was, "Well, she's not really one of us, a clicker trainer as we mean the term. Or she's not one of us yet." Tim Sullivan's remark was more profound. You *can* always tell, but what, exactly, is the difference?

Principle-based training

The good news is that now that we understand so many principles about behavior, and how and why it changes, you don't need natural talent plus 40 years of life lessons to be a highly effective trainer. Principles-based trainers keep the laws of learning clear and in effect from the start. Principles-based trainers also observe their learners and keep emotional signals in mind as useful information about the training progress.

With principles-based training, you can be as creative and effective as only a few geniuses could be in the past, and not just at training animals—at helping with all kinds of human behavior, including your own.

Methods are fun, too

Of course, I'm still interested in methods, though, aren't you? The widening circle of people training from principles leads to invention of more and more new methods, which we gladly share with each other. On the Karen Pryor Academy (KPA) Alumni list I saw some new methods for teaching the obedience scent articles exercise, in which the dog selects, from a pile of predetermined objects, the only one that has its owner's fresh scent on it. (I liked one of the new ideas enough to print it out for evening fun with my elderly poodle.)

Knowledge of principles also lets you weigh one method against another realistically, and include methods with a long tradition of use. (Quick: What's contrary to shaping principles in the grand old method of scenting one article in the pile and tying down all the others so they can't be picked up? Answer below.)

But methods alone aren't enough

Meanwhile, back at the ranch, a new horse is in the corral. A TTouch practitioner, trained by Linda Tellington-Jones, is massaging the front of the horse's shoulder. It's apparent that the horse likes this experience. He stretches out his neck as if she's hit a particularly itchy spot. He waits for more when she stops; in fact he solicits more, by positioning his shoulder closer to her. I don't know the neurological principles that explain this phenomenon. But I can see that the result is very reinforcing, and I wish I knew exactly what she was doing! It's obviously a fine method for winning the confidence of a strange horse.

Thank goodness, I already know how to do that. And not just with horses—with anything with a nervous system. All you need is a few principles—and a click and a treat.

Answer: The shaping procedure benefits from maintaining a high rate of reinforcement for correct responses. The tie-down method offers one correct response opportunity among many incorrect response opportunities, a ratio that is more likely to lead to extinction than to acquisition.

The Shape of Shaping: Some Historical Notes

Born of frustration with a pigeon that wasn't "getting it," shaping was an accidental development that revolutionized—and democratized—training.

Shaping is a concept that many pet owners find hard to grasp. We're used to making animals do things by leading them or pushing them into the behavior we want—and it is hard to believe that there is another way. Common sense tells us that there is no possible way to get an animal to do something it has never done before, doing nothing yourself but reinforcing spontaneous movements.

The term "shaping" was coined by B.F. Skinner to identify building a particular behavior by identifying and marking a series of small steps to achieve it. Shaping allows you to create behavior from scratch without physical control or corrections, by drawing on your animal's natural ability to learn.

Even B.F. Skinner did not start out training animals by capturing and shaping spontaneously offered behavior. Initially, he taught his laboratory animals to press levers and accomplish other tasks by making small changes in the environment: raising the height of a bar in small increments until an animal had to reach higher, or increasing the "stiffness" of a button so a pigeon learned to peck harder. This method was called "successive approximation."

In 1943, while waiting for a government grant to come through, Skinner and two of his graduate students decided to see if they could teach one of their experimental pigeons to bowl in a laboratory on the top floor of a building in Minnesota.

They started by putting the pigeon and a wooden ball in a box rigged with an automatic feeder, planning to trip the feeder when the pigeon swiped at the ball with its beak. But the pigeon did not swipe at the ball as they had hoped, and they grew tired of waiting. Skinner decided to reinforce any movement toward the ball, even just one look toward it. When the pigeon looked in that direction, he clicked the switch, opening the feeder briefly so the pigeon could get a bit of corn.

Skinner later wrote, "The result amazed us. In a few moments, the ball was caroming off the walls of the box as if the pigeon had been a champion squash player." Skinner had made a discovery that astonished even him: It was much easier to shape behavior by tripping the feeder mechanism at the right moments than by changing the environment.

Skinner's daughter, behavior analyst Julie Vargas, PhD, has told me, "His realization at that moment was that if you could do this, you could shape behavior anywhere, in any environment." You did not need to manipulate the learner or build elaborate apparatus. You could just reinforce moves in the right direction.

Skinner named this newly discovered method shaping, to differentiate it from the mechanical process of successive approximation.

Instant gratification

Shaping depends on reinforcing the desired action instantaneously, as it is happening. A key factor in Skinner's early research setting was that the feeders made noise as soon as they were tripped. This click became the conditioned reinforcer that meant food was coming. It was the marker signal that identified the move being reinforced.

Skinner recognized the value of the conditioned reinforcer. For the cameras of *Look* magazine, he trained a dog to jump higher and higher up a wall using a sound and some food; in a popular magazine article in 1951, he recommended the toy cricket or clicker as a good conditioned reinforcer for dog training.

Some people in the behavioral and animal communities have taken to using the word "shaping" to describe any training that increases a response in small increments, even though the response may be generated or elicited by luring, force, verbal instruction, environmental manipulation, or other external pressure, rather than being offered spontaneously. The correct term for these non-spontaneous methods, however, would be successive approximation. Many animal trainers and sports coaches have used successive approximation for years, gradually increasing the height of jumps, the distance of a race, and the heaviness of weights, all to improve performance. The terms "free shaping" and "cold shaping" have arisen as additional ways to identify true shaping, when the animal's volunteered or spontaneous behavior is the key factor in the development of the behavior.

Gifted trainers have also used timely praise and play to reinforce spontaneous behavior and thus develop new kinds of performance without baiting or forcing the movements. The scientific importance of Skinner's discovery was that these principles became generally applicable by any user and in any learning situation, not just by the rare, intuitive, or masterful individual.

Sometimes faster is better

An important characteristic of shaping is the speed with which new responses can develop. This is not a method that requires a lot of practice and repetition.

Often, as Skinner reported with his ball-playing pigeon, complex new behavior can develop in a few minutes. Francis Mechner, PhD, suggests that one explanation for this rapid increase in behavioral topographies is that the marker identifies not only a position—the paw is three inches in the air—but a vector, a movement in a direction. By clicking during the upward movement of the paw, the shaper reinforces not only the current outcome—a three-inch lift—but also the action that is taking place: lifting upward. Reinforcement quickly leads to stronger paw movements and higher lifts, giving the shaper even more and larger behaviors to select.

Birth of clicker training

Keller Breland and his wife Marian, two of the graduate students present at the moment of Skinner's discovery, left psychology to develop a business based on animal training. In the 1960s, they were among the first to carry shaping by use of a marker signal (usually a whistle) into the relatively new world of marine mammal training. Through the late 1980s and early 1990s, after nearly 30 years of development in oceanariums around the world, marker-based shaping spread further, from the marine mammal world into the zoo world and was used for the management of other species by keepers, curators, and consultants, some of whom began their careers as marine mammal trainers.

Over the same period, however, the behavioral research community largely dismissed the importance of the marker signal, focusing instead on the value of the primary reinforcer—usually food—to the learner, whether animal or human. In humans, as when teaching necessary skills to children with developmental deficits, instructors seek cooperation by identifying and using highly prized food items. In shaping behavior in the modern research setting, cooperation is often still guaranteed in animals by increasing hunger, keeping research animals at 85% of normal body weight.

Clicker training, a popular method of training dogs, horses, and other pets using shaping and a marker signal, the clicker, dates to two presentations in May of 1992. One that I organized and led occurred at the annual meeting of the Association for Behavior Analysis in San Francisco and included dog trainer Gary Wilkes (the first to locate and use a commercially available plastic box clicker with dogs), San Diego Zoo curator Gary Priest, and Sea Life Park head trainer Ingrid Kang Shallenberger. That same weekend, Wilkes, Shallenberger, and I presented a seminar for 250 dog trainers outside of San Francisco. In both places we handed out hundreds of free clickers. The widening availability of the Internet fueled the subsequent rapid expansion of the clicker training community.

The shape of things to come

The uses and practices of shaping and its application continue to evolve. In 2001, horse trainer and gymnastics coach Theresa McKeon, together with biochemist and dog trainer Joan Orr and dance teacher Beth Wheeler, began developing the use of the marker signal in teaching physical skills to humans, a practical applications system dubbed TAGteach. As with any emerging technology, new practices ask questions of the basic underlying sciences.

Clicker training and shaping-related studies of both the underlying principles and their applications are underway in behavioral ecology, behavior analysis, sports psychology, and neuroscience. We've covered a lot of ground in a relatively short amount of time.

Lick It!

Sometimes the hardest part about learning to shape behavior is teaching your eyes to look for—and your brain to recognize— the "clickable instant." Here's an excellent behavior to start with.

One of the first exercises in the beginning of the Karen Pryor Academy Dog Trainer Professional program online was to teach your dog to lick its lips, on purpose, for a click and a treat. People who have never done real shaping are *so* accustomed to asking for, causing, or initiating the dog's movements that they sometimes find it very difficult to *see* movements, large or small, that the animal is making on its own.

Seeing the movement

At one of the early ClickerExpos, I was teaching a session on shaping. I offered to show people how to shape the behavior of backing up. To make it as easy as possible for people to see, I brought a Great Dane on stage. I put the owner at the front end of the dog, giving the treats. I stood across the stage behind the dog, clicking for hind leg movements. In short order, the Dane was stepping backwards, then moving backwards a few steps, and then backing up continuously until clicked.

Later that day, a woman stopped me in the hall, introduced herself as a veterinarian, and confessed that the demonstration was very difficult for her to understand.

"I was unable to see the movements," she said.

"You think it's hard to see a Great Dane step backwards? Try it on a Pomeranian," I thought, but I sympathized. Her eyes could see; her brain just couldn't figure out what to focus on.

So, in the KPA course, "lick it" was an exercise in observation, as well as in shaping. The people who found the exercise difficult were often the very people who needed it the most; the more traditional training experience they had, the more of a mental shift they needed to make.

Why the lick?

We chose the lick because it's something all dogs do; in fact, it's something all dogs do often, especially at mealtime or in the hopes of food. Yet you can't *make* the dog do it with a leash, nor do dogs tend to do it when you actually lure with food. (Then a dog is going for the food, so it's sniffing, not licking.) Other reasons for choosing "lick it" were:

- It's brief, so it hones your clicker timing.
- It's harmless, so if it escalates, so what?

My dog never licks

Oh, yes it does. People who are unfamiliar with shaping may feel that they can't click until they see a great big smooch; they don't realize that any glimpse of pink tongue counts. If you're having trouble seeing random licks, however, there is a fast way to get started: butter the dog's nose! Once you've captured a few of the licks the butter elicits, licking should increase enough for you to get the dog licking for a click.

Licking as a stress signal

Some people have doubts about reinforcing licking because it can be a stress signal. However, as long as you are not forcing the action by manipulation or luring (in which case the animal is simply learning to wait for the shove or the lure), the initial cause of an action is not really important.

If the muscles are doing the movement, the click will reinforce the movement. As soon as the behavior becomes operant—hopefully within three or four clicks—any previous "reasons" for doing it are no longer valid. Now the dog's not just licking because it's nervous, or to get the grease off its nose; it's licking to make you click. The behavior itself does not generate an emotion. It's just one more way to earn reinforcement. The likelihood of increasing anxiety by clicking licking is far outweighed by the likelihood of making licking just one more cheery behavior.

We often train a behavior in animals that was initially a symptom of emotion. A horse may rear spontaneously because it is frightened or in an aggressive display. But you can shape rearing, put it on cue, and the horse does it willingly and calmly as a trick or a learned behavior, with no agitation resulting at all. Dolphins and whales use a loud slap of the tail in fear and as a warning signal, but I've trained tons of dolphins to tail-slap on cue without a vestige of emotion attached any longer.

Shaping Your Way to Success

Clicker training has been slow to invade the world of breed shows,
but it offers handlers the potential of hands-free, self-posing dogs,
complete with pricked ears and intent, engaging expressions.
Practitioners rely a lot on micro-shaping.

Marian Breland Bailey's first husband, Keller Breland, left behind this great quote, which I lived by as an oceanarium trainer: "I can train *any* animal to do *any* behavior that the animal is physically and mentally capable of doing." If the penguin could jump across the water, we could train it to jump through a hoop. If the otter could turn a doorknob with its clever paws, we could train it to raise a flag or open a box. If the sea lion could deliberately move its whiskers, we could train it to "smile." If the octopus could squirt water out its siphon, we could train it to make a fountain in the air (and we did).

Dog owners and trainers don't necessarily think that way, though. Go to any dog show and you'll see people using squeakers and other toys to stimulate pricked ears and a lively expression. Why not just train the dog to prick its ears and make a pretty face on cue? You'll see handlers waving and tempting with food, without actually giving it (they call this baiting) to make the dog look lively as the judge walks by. Dogs get bored with that. You'll see handlers physically hauling upward on the leash to keep the dog's head elevated. Why not teach the dog to maintain the head position you want? You'll see handlers manually arranging the dog's feet and then correcting the dog if it dares move. Why not teach the dog to pose itself?

Trouble is, when you try to coax a reaction with some kind of stimulus, over and over, the animal habituates to the stimulus, it stops being meaningful, and you actually get less of the behavior rather than more. When you use physical force or manipulation you may get compliance, but you also build resistance; and if the manipulation is uncomfortable, you build resentment and avoidance.

And Keller was right; there are a host of small behaviors that are impossible to teach and maintain by correction, prompting, or force, but easy to train by shaping and positive reinforcement. That includes everything you want the dog to do in the show ring, including smiling at the judge.

Behaviors you can shape for the breed ring

Teach your dog to prick its ears on cue: Make an unusual sound (squeeze a squeaker once) and capture the pricked ears with a click, treat. Repeat four or five times. Show the sound maker but don't squeeze it, click if the dog pricks its ears even slightly at the sight of the toy; repeat, clicking strong moves and skipping weak moves of the ears. Extend the duration of the pricked ears by shaping, switch the cue to a raised finger, practice in different locations and with distractions.

Correct a bad tail carriage: Here's a tougher show-ring shaping job, but it's by no means impossible—and certainly shaping conscious tail control is better than the owner's surgical solution which is painful, irreversible, and also, of course, illegal. A question came from instructor and handler Vicky Toshach:

> *"I'm having a tough time training a bearded collie not to display a gay tail while being gaited in the ring. [A "gay tail" is a tail that recurves over the dog's back instead of staying at the preferred angle—for bearded collies, curved and low, below the line of the spine.] I had a Parson Russell that had a gay tail when standing but I was able to fix this with clicker training. It was easier because she was standing still, and I just posed the tail where I wanted and clicked. She eventually learned that when she heard the word 'stand,' it also meant to fix her tail! With my client's bearded collie, however, the tail is fine when standing, it's when she is trotting that the tail curls up over her back. One of the methods I tried was using a target stick. I figured that if she was focused on something, she wouldn't be as relaxed and therefore the tail would stay normal. It helped a bit, but still not enough. The owner was contemplating cutting the tendons to the tail. I would sure like to see that avoided."*

Here's the shaping recipe I gave to Vicky. There are multiple steps here, but it's all done in brief sessions with a high rate of reinforcement, so it should go fast and be fun. I have personally reset more than a few show-dog tails with this recipe, including some bearded collies'.

1. Teach the dog to move her tail deliberately, at a standstill. You can use a target stick and touch the tail gently and try to have her move her tail away from the touch. Start by asking the dog to move the tail left or right (not up or down—that's harder) away from the touch.

2. Just moving the tail to the left or right requires the dog to do something consciously that dogs usually do unconsciously. In fact, as soon as the dog discovers that tail moving is the game, the first move will be to wag the tail; it's the only thing they know how to do on purpose. Click that! Then work on just one move, away from the target stick or, free-shape a move to one side, and then to the other, without the target—your choice. That way you don't have to fade the target.

3. When the dog can move the tail consciously, then you can use the target to teach "lower your tail" and "raise your tail." Do this from standing still. Even if you don't want an overly raised or "gay" tail, you should teach the dog to raise the tail for a click as well as lower it. That helps the dog become aware of how it feels to deliberately manage the tail: for them, it must be like you trying to learn to wiggle your ears or cross your eyes. First you need to become aware of the muscles that do that.

4. Put "raise" and "lower" the tail on a verbal cue and fade out the target completely if you have been using one.

5. With someone else leading the dog at a walk (and feeding the treats) introduce the target again as a prompt, for the first couple of clicks, and teach raising and lowering the tail *on cue* with the clicker. This sounds like a lot of trouble but we're talking one- or two-minute sessions, 10 or 20 clicks per session, and in two

sessions the dog should have it figured out. He should have learned the cue and learned how to address the necessary muscles to give a visible response.

6. You do not need to worry about how much the tail is raised or lowered; in fact it's better if you don't. Just click any clear movement without trying to shape a strong movement. You are teaching the dog to make the tail move up or down on cue, *not* to hold it in some particular position. You are literally teaching the dog to think about its tail, and direct its movements consciously, which will be very interesting for the dog.

7. The final stage is to have someone else gait the dog—slowly at first—while the trainer asks for "low tail" if/when the tail starts to go up and over the back. Then click the teeniest effort to move the tail down, even if it's crooked or waving. Stop and jackpot, and try again. If the dog is showing confidence you can start clicking for approaches to the angle you like best.

The "gay" carriage may just be a habit, but also there may be anatomical reasons why the tail curves up and over the back at an extended trot. In either case, it is wise to work by inches and in short sessions. Using rarely used muscles is tiring, and the dog could develop a lame tail! Then the dog would begin to dislike this work and start avoiding it. So follow each session with some fun clicker games or tug or some other play.

Another tip: space out the sessions by three days if you can! In my experience, a much-too-long horseback ride or some other unusual exertion on Sunday makes me feel stiff on Monday but *much* stiffer on Tuesday. If you think for any reason this is hard for the dog, once you get the behavior, resist the temptation to show off and gait the dog over and over with the tail "just right." Wait until the third day to try again! The muscles will get stronger with use.

This is a delightful clicker exercise because it requires free-shaping and good cueing work, and so it is a rewarding challenge and experience for dog and owner too.

Creating a Climate of Abundance

Clicker training improves our ability to notice and reward "good" behavior, so it taps into—and enhances—our generosity in ways that affect the trainer as well as the learner. Just try Karen's "Exercise for Grownups" on page 103. Every click is a gift—"You got it!"—and every reward helps create a "climate of abundance." In one letter, Karen considers the effect this climate of abundance has on families at risk for child abuse. That climate of abundance creates an environment in which gymnasts, bosses, school children, and dragon boat racers all thrive. So do the animals we train. You see it in their attitude and in their eyes—they carry themselves a little prouder.

It takes imagination and empathy, Karen says, to find and use real reinforcers, to discover what hidden reinforcers are undermining our training, and to understand how jackpots work—and how we can make them work for us.

In gymnastics, to help balance after landing the back flip on the beam, it helps to stretch hands, palms down, toward the end of the beam. Here Jenna Moore clicks ("tags") Kayla Jones for keeping her pointing fingers touching on the landing. Photo: Theresa McKeon

A Climate of Abundance

Stinginess, Karen says, is the enemy of clicker training.
If there aren't enough of the right kind of reinforcers, your subject
won't learn—won't even play the clicker game. One of the first
lessons clicker training teaches the trainer, as this piece shows, is
that generosity creates its own karma. That can be a tremendously
liberating lesson for the impoverished, the deprived, or the
Puritanical pet owner to learn.

My friend and colleague Lynn Loar is a social worker specializing in families at risk for child abuse. In one of her programs, she brings several families together weekly for an evening of clicker training, using naïve shelter dogs. The families also play the shaping game with each other—adults clicking and treating children, children clicking and treating grownups—during the course of the session.

On the table is a big bowl of treats for the dogs and another, much bigger bowl of wrapped candy of many different sorts, for the people. As part of learning to click when they are playing the shaping game, the families must also learn to give treats to the dogs and candies to the other people. Adults and children also earn clicks and candy from Lynn or her helpers. Lynn and her staff might click and treat a person who was previously talking or turning away but is now watching the proceedings, or when they see a smile or an instant of warmth, such as an adult putting a candy into a child's hand instead of just tossing it.

"People have a history of deprivation," Lynn says. "We all feel deprived sometimes. For the adults in these families, it's hard to give anything away, even a smile or a touch, when they never got anything themselves. That big bowl of candy on the table is obviously *far* more than we need, far more than we're going to use. There is plenty. There is more than plenty. That bowl in the room creates a climate of abundance, a climate in which it's all right to be generous to a child, to feel interest in a dog, to give treats away instead of hanging onto them. The second week, the first thing people do when they come into the room is look to see if the bowl of candy is still there." And it always is. The very existence of that lavish supply creates a climate of abundance. By the sixth week, families leave the session holding hands, laughing, and chatting with each other. The whole style of interaction has been transformed for the better.

An undercurrent of stinginess

In spite of holidays, such as Christmas, which are given over to generosity and to quantities of presents, food, and drink, there's an undercurrent of frugality in our culture, almost of stinginess. It's the enemy of clicker training.

New clicker training students often seem to feel that giving a dog 50 or 60 treats in one 10-minute session is just plain wrong. It's evil. Once, about 20 years ago, I was giving a seminar in New England and there was a set of weave poles in the room. That night, I thought of how I might train weave poles by back-chaining. So, the next day I found a lively Jack Russell, ascertained that he had never been exposed to any agility training, had him crated for a couple of hours so he'd be hungry and restless, and brought him up on stage. I trained a behavior (now called the macaroni, I'm told), put it on cue, back-chained it, and in about 10 minutes had this little dog running the whole line of poles. Yay—that worked!

But to this day I can see the owner's sour face. She didn't care about her dog's success; she was disgusted by the number of treats he got, enough to make him sick, she was sure. (It didn't—I watched. It just made him very interested in weave poles.)

Start small, create a chain of plenty

Maybe we can create a climate of abundance, not by overeating, not by overspending, but by thinking it through. Small abundances can make that feeling happen. I've done it by dividing one bunch of flowers into three or four and putting little arrangements all over the place. Or by baking three pecan pies instead of one, and giving two away. Who learns about being free-handed with reinforcers, instead of fearful and stingy, by doing this? Not just the recipient. You.

Have You Ever Been Right, 47 Times?

In our often-frugal society, we tend to be stingy with
compliments and positive feedback. This story amply illustrates
what a barrage of positive reinforcement can do beyond just
teaching you a skill (or, in this case, extinguishing a bad habit).
It affects your whole being.

When you are training people who are training their dogs, it helps to
have two "clickers," each making a distinct noise so you can mark the
human while she marks the dog. (Two distinct marker sounds also make
it easier to clicker train multiple dogs, or multiple pets of any kind. Each
dog learns to distinguish "its" click easily.) I found a toy that made a "ping"
sound that was perfect for my purposes.

Training the handler and the dog at the same time

The first time I personally used two kinds of clicks was on stage at
ClickerExpo in 2005. I wanted to explore a new way to get rid of the pernicious leash-jerking that seems to be so prevalent these days, an aversive
being used routinely to signal the dog to come along when you move. (For
more on this pet peeve of mine, see "Hidden Aversives," page 159.)

In a session on cues and cueing, I asked people in the audience to walk
down the aisles with their dogs so I could select two or three audience
members who had this leash-jerking habit. I brought them up on stage and
narrowed my choice down to one, a woman with a nice, polite standard
poodle and a habitual leash jerk.

I gave her these instructions: "Hold the leash in your right hand" (so jerking wouldn't be as easy to do). "When I say 'go,' tell your dog 'let's go,' and start to move forward down the length of the stage and back. When you hear the click, stop and give your dog a treat."

Then we began. The stage was small, and the pet owner had to make U-turns at each end, so that took some extra coaching to avoid pulling the dog through the turns, but we managed that.

With my left hand, I clicked the dog at least once on each straightaway if it was in heel position (it always was), and I clicked if it stayed with her on the U-turns. Meanwhile, each time the handler stopped without jerking the leash, I played the ping sound on the toy in my right hand. And each time I said "go," if she spoke to the dog instead of jerking on the leash as she started forward, I pinged her again. I had a stash of wrapped chocolates in my right hand. After every ping I moved into her path and handed her a chocolate. Pretty soon she had a pocketful.

Without the incessant leash jerking, the dog actually knew how to heel very well, and was stopping, starting, and turning when his owner did. He was enjoying all the clicks and treats and working attentively. The handler, meanwhile, had earned about 10 pings and was beginning to do a nice, co-ordinated job of stopping on the click, going on my "go" cue, and speaking to the dog before moving, almost every time, with absolutely no tightening of the leash.

On about the 10th or 12th ping, the handler took her chocolate, pocketed it, then suddenly looked at the noisemaker in my hand, pointed at it, and said "That's for *me!*" I nodded and smiled, "You've got it!" And she did. We'd finished the training, as far as I was concerned.

When she left the stage, the dog was glued to her side. She spoke to the dog before going down the steps, and she went back to her seat without jerking the leash once.

The exercise had worked. I was pleased. And it had an unexpected result, too. Later in the day, out in the hall, this handler and her dog walked past me and she stopped (without jerking the leash!) to talk. She told me that the five minutes or so of training had been a tremendous experience for her. "I'm always such a klutz," she said. "I never get things right. But this, I got this right, and it felt so good! I feel completely different. I was a success!" Yes, she was. I'm sure I was grinning. Her dog was laughing, too. The pet owner had experienced, in real time, learning a new skill with a very high rate of reinforcement. That was huge. She was standing up straighter, and her face was relaxed; she was smiling. I had put her on the spot in front of 250 people and that didn't matter a bit—she felt great.

It put me in mind of a video clip TAGteacher Theresa McKeon shows. She had been teaching cheerleading moves to some young girls, perhaps 10- to 12-year-olds, using the clicker, or TAG. When the lesson was over, she asked the children about their experience.

"How many TAGs did you earn?" Theresa asked one girl.

"Forty-seven," the child said promptly. Others chimed in with their numbers—they counted; they knew.

"Is that so?" asked Theresa. "How many times in your life before now have you been told you were right, 47 times in one day?"

"Never," said the child, giggling. Everyone else giggled too. What a weird idea.

That's right. We get told when we're wrong. We don't hear about it when we're right, certainly not 47 times. And just having that experience once can be life-changing. Even if you're no longer 10 years old.

An Exercise for Grownups: Changing your View

Especially in a world where criticism is the norm, we don't get a
chance to exercise calling out the positive—and often we don't even
see it. Karen encourages us to practice—and enjoy the results.

Here's an exercise anyone can try.

- During the day, make a point of noticing something someone
  else is doing that you like—someone at work, someone at home,
  a stranger even. It need not be something unusual. It can be
  something you already expect him or her to do anyway.

- At the end of the day, find time to tell the person that he or she
  did that thing right. Avoid the word "I" as in, "I liked the way
  you..."

  o "I" is all about you, not about the behavior. Instead, just
  name the behavior.

  o "It's good that you finished your homework." "You handled
  that phone call well." "The client report is done; that's great!"
  "The kitchen's all cleaned up; that's so nice."

- With kids, learn to watch for one or two things that were notable.
  Don't turn it into a big deal; just identify something the child did
  right.

  o "Hey, you got down to dinner on time—thanks." "Really
  appreciated that you fed the dog on your own." "It was great
  to hear you reading to your sister."

- Don't make things up afterward; really take notice during the day. What you are trying to train here is not the kids, but your own observant eye.

- Find a quiet time at the end of the day, bedtime for instance, to tell what you saw. A thank you at the time is good, but a thank you during a quiet time together can really sink in.

A slow shift, a long-term gain

This new effort may feel weird. You may feel self-conscious. That tells you that you need practice noticing and discussing things you like. (Do you have more experience at noticing and discussing things you don't like? That's true for most of us.)

Don't expect any particular response; this may be new for the other person, too, and may take some getting used to. Just keep it up. The change in the behavior of the recipients of these observations will be rapid and obvious. The change in your own habits and observation skills may take longer, but it will be subtle and profound.

Jackpots: Hitting It Big

If there's any group of people who can teach us about reinforcement, it's casino owners. Their livelihoods depend on not only engaging the SEEKING system full-bore but also on making winning the biggest, bestest experience possible. No wonder Karen uses the slot machine as the model for a jackpot. It's not what you probably think it is.

Jackpots!

A good thing? A bad thing? A nonexistent thing?

Let me tell you what I mean by "jackpot." I mean exactly what the casinos mean: a surprisingly big reinforcer, delivered contingently. The key is in the word "contingent." To reinforce a particular behavior, a jackpot has to appear, and be perceived by your learner, while the learner is doing that particular thing you want. Not afterwards.

If you click, and then deliver the treat afterwards, an especially large, numerous, or wonderful treat is no different from any other treat in terms of its ability to reinforce behavior. Good treats or big treats may make the learner more interested in the training and work in a general way, but they're not specifically informative. The association is Pavlovian. Click means treat is coming. If the treat is sometimes a kibble and sometimes chicken, sometimes small and sometimes huge, that's fine. It keeps your clicker nice and strong, but it doesn't tell the animal anything different about the behavior. So that's not what I mean by a jackpot.

People often assume that a jackpot is any unusually large reward. An obedience competitor tells me, "I use jackpots; we go through the ring work, and then back at the crate I have six or eight wonderful treats waiting, and he knows that ahead of time." Well, maybe he does, and maybe he doesn't; hard to say what a dog "knows." And what would you say those treats reinforce: the work in the ring? Getting back to the crate, would be my guess. That's not a jackpot. That's just a windfall, an unusually large treat, associated with nothing in particular. Nice, but…

A parrot owner tells me, "I use jackpots. When he's done something especially good I make a big fuss and give him six pieces of fruit, one at a time, instead of the usual one piece." Same deal. What does that tell the parrot? "Sometimes my person gives me more fruit than other times." Nice, but… <shrug>

Several people have posted that they tested the "jackpot" theory, usually with a setup such as this: You divide treats up into small and large amounts, and then reward one behavior with the click followed by small amounts, and another behavior with the click followed by large amounts. Usually there's no discernable difference between the results. Both behaviors are maintained at whatever level you're clicking for. Ergo, jackpots don't work?

No, because that's not a jackpot. To me, a jackpot is only a jackpot if it's delivered in such a way that it functions as an *event marker*, identifying the wonderful act as it is happening, and also functions as an unusual primary reinforcer, making that behavior more likely to happen again.

A real jackpot should be such a surprise, and happen so suddenly, that it is actually startling. Think about the casino slot machine jackpot. Ever won one? You're playing away, shoving in quarters, pulling the handle, getting close to a win now and then, or getting two or three quarters back which you feed into the machine again, and just when you're thinking of quitting and getting something to eat, you pull the handle, watch the things spinning around, and WHAMMO! Bells ring, the machine lights up and flashes (those are clicks, of course), and, with a rattle and roar, a hundred quarters pour out of the machine. What a surprise!

What were you doing at the time? Watching the wheels in the machine spinning around the way you made them do by pulling the handle. What behavior was reinforced? Watching the wheels roll around while hoping to win.

To repeat this experience you have to get more quarters, and pull the handle again to make the wheels spin so you can watch them in hopes of another jackpot. It's a little behavior chain. To maintain the chain, only the last behavior, watching the wheels, needs to be reinforced, and then only very sporadically (you're on a long variable ratio schedule, natch, which can be addicting). Watch the people in a casino intently staring at the screens of the slot machines they are "playing," hour after hour after hour.

OK. Now imagine if what happens is you get the wheels spinning, and they come to a halt, and there is a sign saying, you win. But you get your winnings non-contingently. When you go back to your room, you discover 100 quarters on your bed.

The amount is the same. But what is it actually reinforcing? Maybe you feel you'd like to come back to this hotel again. Maybe you'll drop in on your room more often. But you may not feel at all like going right back and making the slot machine work some more. You might decide that spending the money in the gift shop down the street would be more fun. Believe me, casinos never make that mistake. Their jackpots are not only extremely noticeable (to you and everyone else), they are absolutely contingent on what the casino would like you to keep doing: Putting your money in slot machines so you can watch those wheels go around.

So, in working with an animal, I would never use larger quantities in the hopes of maintaining behavior that's already trained. I would only use an unusually large quantity as a true jackpot, that is, delivered all at once and contingently, to both mark and strongly reinforce the first occurrence of a rare behavior, or the first achieving of a difficult move.

If I wanted to give a dolphin a jackpot the first time it swam through a really scary hoop, I'd blow the whistle and upend the fish bucket simultaneously: it's raining fish! I have given my dogs jackpots for coming when called from long distances for the first time.

In *Don't Shoot the Dog* (pages 11-13, revised edition) I wrote about jackpots, giving what I still think is a good example of the jackpot: the tactic of the horse trainer I mentioned on page 11. He worked with American Saddlebreds. These horses are taught a somewhat unnatural, four-beat gait called the "rack." They are bred to do it, but they also may have to be helped by the rider at first. When working with a young horse, it was this trainer's habit, when the colt "first hit a lick of the rack," to jump off him immediately, strip off the saddle and bridle, and turn him loose in the ring. Now that is a jackpot. It's negative reinforcement, freeing the colt from the demands of the rider, but reinforcement, all the same; and the fact that it was a completely new and thus memorable event makes it momentous. I think it also facilitates remembering the brand-new movement that made it happen.

Remember, though, that no one is always right. We all keep learning things all the time, revising and adding to what we knew before. That's one of the joys of science and a valuable phenomenon in the clicker training world.

Looking back at the jackpot section in *Don't Shoot the Dog,* now 20 years after it was written and 6 years after the new 1999 edition, I think that I failed to differentiate between jackpots as I see them and another tool altogether: the non-contingent reward.

A non-contingent reward is also something you get by surprise, but it is not associated with any particular behavior. One example in the book was the two free fish we gave to a discouraged dolphin, which perked her up and set her to trying to earn reinforcement again. Another example in the book was the ticket for 10 free riding lessons that my parents bought me when, at 16, I was behaving poorly for weeks on end. It instantly corrected my bad mood. I included these as jackpots, but they were not; they were both examples of a non-contingent reward. The most powerful use of a non-contingent reward is to counteract the effects of an extinction curve. I know the dolphin in question was undergoing extinction of a bunch of operant behaviors; probably the sulky teenager was, too. Getting the news that good things are still available re-engaged the SEEKING system.

Like the jackpot, a non-contingent reward is a tool to use rarely. And, like a jackpot, if it is going to work, you only need to do it once.

Clicker Training for Gymnasts

The setup of a clicker-based gymnastic class, Karen
shows us, is a perfect way to bathe students in positive
reinforcement, and they respond accordingly.

In February, 2004, I traveled to Charlotte, North Carolina, to spend a
few days with gymnastics coach (and ClickerExpo faculty member) Theresa
McKeon, her husband Brian, and their kids. While the purpose of my
trip was to visit with Theresa and her clicker trained gymnastics students,
after single-digit temperatures in Boston during most of January, the balmy
early spring of Charlotte was a blessing. Early crocuses were appearing, and
a few cardinals and robins starting to sing a little. Theresa's horses were
lying on their sides in the pastures, soaking up the sun.

We spent one happy afternoon in the sun playing with a dear old horse
that Theresa is clicker training, but most of our time was spent in the
gym. The gym is a huge structure, big enough to hold 200 to 300 girls at
once. The girls range from about seven years old to high school seniors.
They were divided into groups, working in different areas and on different
skills. I was enchanted by the sight of all those children flipping themselves
around in the air like spinner dolphins.

This is a serious matter, however, not just recreation. Many of the girls
start competing at state and regional levels when they are no more than
nine years old, and this organization, with coaches like Theresa, produces
national champions. The goal is not just exercise, or fun, or medals and
glory. It's college scholarships, and big ones, too—something really worth
working for.

The clicker is an ideal tool to identify correct movement, especially in the air. Theresa and her scientist colleague Joan Orr call it "TAG" teaching—Teaching with Acoustical Guidance. It's clicker training, of course (the clicker is the acoustical marker), but giving it a different name helps parents get past the idea that coaches are using "dog training" with their kids.

And boy, does it work (of course). I watched Theresa coaching a bunch of teenaged girls through a workout on the uneven parallel bars. The girls had already learned many parts of the exercise via shaping and reinforcement, one element at a time. Now they could perform the whole exercise: swing up to crouch on top of the low bar, jump from the low bar to catch the high bar, swing under the high bar and up into the air again, let go, and do a back somersault in the air, landing on their feet on a pile of mats. (This part of the maneuver is called a fly-away dismount.)

One by one, they tried it. All of them got up on top of the first bar all right, and caught the second bar. But most of them flubbed the dismount and landing, falling backward out of the somersault onto their rumps instead of landing on their feet.

In each maneuver, Theresa looks for what she calls the "tag points"—the place where she can teach the correct action that leads to success. In this case, she said, kids were losing energy in the jump from the low bar to the high bar. The girls were letting their legs flop in the air, which pulls the body out of alignment and leads to a loss of momentum.

Traditionally, a coach might deal with this by ordering the girls to try harder as they start the maneuver, and by scolding for bad landings after they happen. And of course, the coach is usually yelling, since the gym is pretty noisy—so she may sound angry even if she isn't.

Theresa called the girls together and said quietly, "The tag point is this: legs together as you jump to the high bar." The girls nodded solemnly, lined up, and began, one-by-one, running for the low bar, catching it, swinging forward, back, forward, and then up into the air to land feet-first on the low bar. From this position, they instantly jumped through the air to the high bar (this move is called a "kip, cast, squat on"). As each girl flew through the air, her legs were straight and together, her toes pointed: Click! Sometimes Theresa accompanied the click with a shout of praise, "Nice! Legs together, good!"

Each girl then caught the high bar, swung under it, let go, flipped into a somersault, went all the way around—and landed on her feet! For some it was their very first successful fly-away dismount. They looked not only pleased, but rather surprised. "Oh, wow! Did you see me?"

As an expert gymnast herself, Theresa knew that getting into the right position would add enough energy to the swing to bring the girls over and onto their feet. That's the coach's expertise.

What the clicker does is give the coach a way to talk to the muscles, not the mind. The sound is like a snapshot, a picture of the action. For the tagged child, it identifies exactly what the right move feels like, from top to toe, and *that* enables her to do it again. Instead of practicing and making mistakes, and getting scolded, and making more mistakes until she gets it right, she can get it right once, and then practice that!

For all the watching girls, it identifies what the right move looks like, and when it has to happen, so they learn something, too. In fact, once the girls all get the idea, they can take turns clicking each other for specific tag points. It's a great game, and mutually reinforcing. Doesn't that make sense?

They think so. Here's what some fourth- and fifth-grade girls told me:

"Coaches tell you what you did wrong. The clicker tells you what you did right."

"When you don't get a tag, you know to change what you're doing until you do get it."

"When you fix it yourself, you remember it better."

"We can tag each other, and that helps a lot. The more you are the teacher, the more you can get it right yourself."

"It's a fun sound."

"Clickers never yell."

Click!

The Mean Boss

Karen shares the secret for reforming the naggers, whiners, blamers, and teasers in your life. It's all about when you give them the attention they crave.

Once at a conference I shared a breakfast table with a woman who was an executive in an oil company. Finding out what I do for a living, she expressed a firm belief in reinforcement, and told me the following story.

Click the boss

She had a mean boss. He was brutal: blaming, finding fault, belittling, angry. Everyone, including the woman who was telling me the story, was alarmed to see him come into their office and glad to see him go.

"This makes no sense at all," she told herself. So she developed a tactic. When he came to talk to her, if his tone was pleasant, she looked at him. If his words became unpleasant, she looked away. If the voice became normal again, she looked back. When the boss stayed calm, she offered him candy.

By and by, he got nicer. Poor man, she thought. Probably no one is ever nice to him since he preempts any conversation with angry behavior.

The boss began to be calmer with others, so much so that people were commenting that he seemed to be in a good mood lately. So then the woman passed on her "secret"—look away and look back. Everyone tried this, and the whole climate of the office changed. Now she's happy, and so is everyone else, a lot more work gets done, and people even stay late sometimes.

Message from Mimi

I was telling my upstairs neighbor this story, and she said, "That's what Mimi does." Mimi is a petite cream-colored Burmese cat, now quite elderly, that used to belong to me. She stayed with this neighbor when I travelled. Eventually, since it's a lot less boring upstairs where several people live than it is in my first-floor apartment with just me, Mimi moved upstairs permanently and has lived there ever since.

But remember, she was once my cat, so Mimi is a clicker cat and knows about behavior.

My neighbor went on to explain:

"Mimi looks away when she doesn't like what I'm doing. And when she likes it, she looks right at me with this warm, bright-eyed, eager expression. I'd do anything to earn that expression! So she's taught me what she likes and what she doesn't like. In fact she's taught everyone in the house. Take Robert, for example. Mimi's sitting on his lap and he's scratching her chest. After a while, she looks away. Oops! Robert stops scratching. 'She told me that was enough scratching; she looked away.' Mimi doesn't want to leave, because he's warm and cozy and she likes sitting on his lap, but she made her thoughts clear."

The cat knows what works for her!

Try this at home, at work, in the neighborhood, at the market...

Do you have a nagger, a whiner, a blamer, or a teaser in your life? Those behaviors are commonplace because they work. You give people who behave this way attention, even in just trying to stop them. How about trying what my breakfast companion and Mimi knew? Looking away, by itself, doesn't do the trick. It's the looking *back* (perhaps with Mimi the cat's smiling eyes?) in the instant the unpleasant tone stops that makes it work.

Paying Kids to Learn

Could clicker-training principles improve kids'
performance in school? You bet!

In 2010, *TIME* magazine ran a cover story about paying kids cash to get better grades.

The objections to cash "rewards" for schooling have been around for a long time and can lead to tremendous political uproar. There are moral objections—children should do what's expected of them without reward. There are philosophical, theoretical, religious, and, of course, financial objections.

Well, this fellow at Harvard, economist Roland Fryer, Jr., decided the first thing to do was to find out if paying kids to do better in school actually worked or not. Forget all the existing studies and opinions. Forget those specific schools where reinforcers, large and small, are built into the system. According to *TIME,* Dr. Fryer "did something education researchers almost never do: he ran a randomized experiment." (Just think about *that* for a minute. They opine stuff and put it into the schools, and they don't *test* it?)

Anyway, Fryer persuaded four cities—New York, Washington, D.C., Chicago, and Dallas—to set up ways to pay some groups of kids to learn (while others just did the usual learning). The experiment involved 18,000 kids and a total of $6.3 million in payouts.

Fryer left the design of the programs up to the cities; he let them pick whatever they thought would work. The results, which he shared exclusively with *TIME,* "represent the largest study of financial incentives in the classroom and one of the more rigorous studies ever on anything in education policy."

City planning

New York set up a program to pay fourth- through seventh-grade children for their test grades during the school year. For great results you could get as much as $50. The money went right into a savings account.

Chicago also paid for test scores during the year. Good scores could earn up to $2,000 per year, half of which went into a savings account payable on graduation.

Washington, D.C. had a complicated system in which high school students were paid $100 every two weeks by getting perfect marks in five different areas, including attendance and good behavior.

Dallas kept it simple. Second-graders got $2 every time they read a book and passed a little computer test on it.

Then, to see if their scores had improved over those of control groups who got no money, the kids all took the national MCAS tests at the end of the year.

What happened

In New York and Chicago, attendance improved, morale improved, grades improved, and the kids liked the program—but the MCAS test score improvement was zero. Nil. None. Nada.

Washington, D.C. showed distinct improvement in general behavior and, presumably as a result, some improvement in reading scores, enough so that the chancellor was thrilled and extended the program after the experiment was over.

And 85% of the Dallas second-graders improved their reading the equivalent of five full months of extra schooling, and continued to improve the year after that.

Why the differences? Clicker trainers could tell them...

In Chicago and New York, the event being reinforced—grades on tests—was an end result, not a behavior in itself. The money, too, accumulating in savings or paid out at graduation, was seriously delayed, functioning as a positive experience but not necessarily a reinforcer. Sort of a lure; gets you hopeful and moving, and in a good mood, but doesn't actually teach you much. Kids loved the program, and wanted to earn more; they just didn't know how.

In D.C., most of the five behaviors (coming to school, not fighting) were things that the kids could control and could offer deliberately. $100 every two weeks was frequent enough to actually reinforce better behavior, and a global change in behavior enabled everybody to learn more. Standardized test scores in reading went up about three months' worth, even though nothing else in the teaching or school changed.

And in Dallas? The behavior was a clear-cut operant behavior the children could already do: read a book and answer a quiz on screen. The payoff was connected to the task and was therefore a reinforcer. MCAS reading scores improved by five months. It was as if the kids had had another half year of schooling. And it cost Dallas about $14 a kid.

What of the 15% of Dallas children who did not earn pay and did not get better? Perhaps they were the ones that couldn't really read yet, or at least not in English. They couldn't earn reinforcement because they just didn't have the behavior. *TIME* thought so, too.

My take on it

Fryer is reported as saying he doesn't really know why it worked best in Dallas, or why 15% of the Dallas kids didn't learn. He does know, however, that they definitely have an answer to the question, does money work. Done right, cash can make a huge difference.

How exhausting—four cities, 6 million dollars, 18,000 kids—and only one school system came up with an operant behavior and a timely reinforcer. And no one noticed those fundamental facts. Makes you want to laugh and cry at the same time, doesn't it?

Well, good for Dr. Fryer and *TIME* magazine. Maybe *someone* besides us clicker trainers will read the story and say, "Oh. I see why that worked. Let's get it going in our school."

Our town. Our city. Our state. Our planet.

Clicking below the Surface

A Taiwanese firefighter proves that you can TAGteach
anywhere—with startling results.

In 2008, I went out to Seattle to get to know Karen Pryor Academy's first class made up exclusively of international students: one from Hong Kong, two from Taiwan, two from Israel, one from Finland. Terry Ryan's Legacy Canine School was a wonderful location for this class. Her building is beautiful and well designed, and it's located on the spectacular Olympic Peninsula, just about a two-hour drive from Seattle. The school has the sea on one side, lush farms and gardens in the middle, sun all day, every day, and snow-capped mountains all around. By the time I had arrived, the students had completed their online lessons and at-home exercises and were ready for an intensive hands-on program under teacher Terry Ryan.

Do you know how to paddle a dragon boat?

Even before the hands-on class in Seattle had started, one student, also named Ryan, had put his new TAGteach knowledge to work. Ryan is a Taiwan firefighter, head trainer for their canine search and rescue team, and, of course, a clicker trainer. He's also part of a competitive crew in the major sport of dragon boat racing. The biggest races are held every year in Hong Kong, but races are held in many other cities around the world, including in Boston.

Dragon boats are 30 meters long and carry two lines of paddlers. Paddlers do not pull their paddles through the water in an arc, as we do when canoeing. (If you use your arms that way, you get tired fast!) Instead, paddlers must use a more powerful stroke—thrusting the paddle straight down and moving it straight back horizontally, synchronized with the other paddlers, of course. You have to make the move with your back—with your whole body, really—twisting from the hips each time. It's hard for people to learn, and it takes forever to train the whole crew to do it right.

Clicker trainer overboard—and then success!

Back home in Taiwan, Ryan started with a new crew and a new coach. He introduced the clicker, and tagged the crew for each aspect of the paddling move. Bingo—everyone began to get the idea.

Their coach took the crew to a swimming pool and had the paddlers line up on the side of the pool and paddle. Next, the coach put on a mask, went in the water, and ducked underwater to look at the paddle moves from that perspective—to make sure the paddlers were paddling straight. Of course, the coach had to come up to the surface to give his report and to announce if the effort and results were satisfactory—a delay. The coach learned something, but the paddlers? Not so much.

So Ryan gave the coach a clicker and some instructions: Go under water, look at the paddling, and stick your arm over your head and into the air to click when the paddles are moving correctly. Again, bingo! Soon everyone was paddling just right.

Then came the race. The other teams were not only more experienced, but bigger and stronger and fitter. "Our guys are skinny," Ryan said.

Guest what? Ryan's crew won! Champions, the first time out. Click!

The Art of the Possible

If you think it's hopeless to teach your dog to ignore
such reinforcers as teasing squirrels or treats strewn on the
ground, the real challenge, Karen says, is rethinking the
art of the possible.

You can view something your learner wants to do very badly as "competition" for the treats you carry, a behavior flouting your level of control—that would be the intuitive and historical way. Or, you can view the coveted activity as a powerful potential reinforcer, and set things up so your learner gets paid with a chance to enjoy that reinforcer for behavior that you want—which, in fact, may include ignoring that reinforcer. That's the science of it. Chris Puls's shaping of "leave it" in a beagle was a good example:

> *I just spent an hour working with my beagle, using the training steps I laid out for "leave it." He was completely off leash the whole time. He started out having difficulty looking away for more than one second from the treat held in my hand out to my side (eye contact game). By the end of the session, he was walking over and around treats on the floor! While off leash! And he was doing "leave it" on treats I deliberately dropped on the floor!*
>
> *Working on getting the eye contact when I presented the hand with the treat in it took half of the session, especially to get the eye contact solid for 10+ seconds while my hand with the treat was moving a bit—and regardless of where my hand was placed in relation to him.*

After the long time of working on getting him to look away from the treat in my hand, things progressed a bit faster. When I switched to the treats in a Ball jar lid in my hand, though, I had to start over and have my thumb over the treats. He went through the steps faster this time than he did with just the eye-contact first steps.

Soon I was able to put the treats in the lid right near him after I said "leave it" and he was looking right at me, even though they weren't protected by my thumb. We took a break here (so I could cut up more treats) and played a short game of fetch the toy, but he kept running right back to the training area with his tail wagging so I knew he was still ready to learn.

I moved to a few different places/positions with the lid in my hand but when I stood, it was really hurting my back to bend over (to get the lid close to the floor where it would end up). So I taped the Ball jar lid onto the end of a dowel so I could offer it low to the ground and remain standing.

Then I did some reps where I held the lid and set it on the ground. I kept my hand near the stick that was still attached just in case. When I felt he was ready/reliable I removed the stick I had taped the lid onto. He would look at the lid when I put it down and immediately give/hold eye contact. Yea!

I upped the criteria by asking him to come past the lid to get the reward/treat after I clicked. I was able to get quite a distance away and he was still not going for the treats, even though there was no way I could have protected them. This is farther than we have ever gotten! And his behavior was a lot more reliable than ever before.

Next, I removed the rim of the lid, so the top part was flat on the floor. No problem. I guess it still looked enough like the lid even though it was now just a flat disk on the floor with the treats on top. Problems arose when I dumped the treats off that lid and onto the floor. I didn't break down the criteria enough, and he scarfed up all the food.

We tried again with loose food on the floor while I was much closer and able to move a foot in to protect the food (only had to move my foot a bit toward the food when I thought he was going to go for it and, after that, he didn't try to get it again). It was very clear that he really wanted to! But I was also comfortable trusting him to respond to the behavior that had been the most rewarded: eye contact.

Imagining what's possible

The crossover trainer—and I am one, too, after all—often comes to clicker training with a background where success has included correction. For years, in seminars and classrooms I have been faced with the telling phrase, "I don't see how you could possibly..." followed by an example, often drawn from extensive expert experience.

"I don't see how you could possibly train x to do y under circumstance z, without punishment," the person says. The declaration is often made as a kind of ultimatum or challenge. "There! See that? There's no clicker answer to that!"

The key is not in the x-y-z of the example, but in the opening phrase, "I don't see how..." The person really doesn't see how another way is possible, and that's the only issue here. Nor should the person be expected to; it took Skinner *et al.* a long time to come up with the principles underlying operant conditioning, and it has taken all of us dog trainers a decade or more to devise, use, and teach how to apply the principles effectively in challenging situations.

Meanwhile, the challenger deserves empathy. This situation, when new information is replacing old information, resembles the process called extinction, in which previously reinforced behavior is no longer reinforced. Extinction is a very unpleasant process for any organism. The experience often arouses anger, even in lab animals.

A squirrelly reward?

You can help people through the extinction curve by explaining and teaching the shaping/cueing approach to managing behavior around strong attractions. One does this, of course, by incorporating the attractions into the shaping plan.

Dealing with squirrel chasing in the woods is just a shaping staircase; if you want to do it, it can be done, but it involves a lot of steps. For me, that's too much like work. My practical solution is a mix of training and management. The backyard is fenced, and there the dogs can bark and chase squirrels all they want. Outside the front door, on the sidewalk, we enjoy a shaped behavior of stalking squirrels, with an occasional brief "chase" reinforcer. In the woods, my poodle, whose lust for squirrels is mitigated by his general timidity, can be off-leash, because he was quite easily shaped to come when called, even from squirrels. My 17-year-old border terrier, however, stays on-leash in the woods. From her standpoint, it's a lot better than no woods at all.

I now live on a tree-lined Boston street near a city golf course with lots of oak trees. We have a huge squirrel population. My timid miniature poodle Misha was afraid of squirrels when he first saw them, but my border terrier Twitchett explained that you have to chase every squirrel you see; all dogs know that. So he learned. One day a squirrel ran across the sidewalk in front of us and Misha leapt at it and nearly jerked my arm off.

So I devised a training plan. A day or two later, on our afternoon stroll, a squirrel showed up on the sidewalk ahead of us. Before the dogs saw it I got out my clicker and began walking very, very slowly: step, click, stop, treat. Step step, click, stop, treat. The squirrel ignored this slow approach. It was near the base of the tree, so when we were about 15 feet from the squirrel, I dropped the leashes and ran toward the squirrel. Of course the dogs dashed forward too, and I cheered them on. The squirrel leaped onto the tree and disappeared up the far side. Thrilled by the apparent near-success, tail wagging furiously, Twitchett put her paws on the tree trunk and looked up into the branches, while Misha, puzzled, looked in all directions.

That was all the reinforcement needed. From then on when we saw a squirrel the neighbors were treated to the sight of a grandmother and two smallish dogs, all crouched, intent, creeping slowly down the sidewalk, stalking squirrels. Every time, 15 feet from success, I dropped the leashes and we all chased the squirrel up the tree. No more clicks, no more treats. Just the thrill of the hunt. The neighbors, I am told, assumed I was training the dogs to kill squirrels. The dogs felt sure that one day they would be successful. We all enjoyed the last-minute rush, possibly even the squirrels, since in the past I have witnessed squirrels similarly teasing dogs on purpose. And neither dog ever yanked my shoulder off again.

Can a Punisher Also Be a Reinforcer?

Most of us view those who celebrate New Year's Day with a Polar Bear swim as bonkers. Karen takes another look at the behavior and who defines a reward or a punishment.

On New Year's Day my otherwise wonderfully sane son-in-law took his family to the beach and went for a swim. Big deal, right? But this is Boston. The temperature was 11 degrees at my house that morning. We had just had a big snowstorm and were expecting another. Even the ocean was practically frozen.

Karl and a friend have been celebrating New Year's Day with a Polar Bear swim for some years now. They like to do it. It's my idea of extreme punishment, but their idea of fun.

Would non-human animals ever do something like that? KPA graduate and Certified Training Partner (CTP) Dan De La Rosa asked me, "Can a punisher also be a reinforcer?"

Dan says that Schutzhund working dogs take stick hits from the target person they are attacking and come back fighting for more. It seems that they do not see the stick hits as punishment. Some world-class trainers describe these dogs as adrenalin addicts.

Choose your words carefully

The short answer to Dan's question: don't equate an aversive stimulus with a punishment. A punishment, technically, is anything that shortens or stops a behavior. If someone hit you or me with a stick, we might do less of whatever we did that triggered an adversary to strike at us; however, a fighter might do more. I shrink at the very thought of swimming in freezing water; Karl jumps in, once a year.

An aversive stimulus is not necessarily punishing; it depends on circumstances and on the recipient. Rain is punishing to cats, who might respond by going inside; reinforcing to ducks, who might respond by going outside; and a matter of indifference to cows, who stay where they are. And circumstances can change one's view of what's punishing. We primates generally seek shelter in a downpour, but Gene Kelly, having just fallen in love, famously sang and danced in the rain.

We need to separate the "thing"—the cookie or the stick or the click or the cold water or whatever—from its outcome. What defines its function is not what it looks like to common sense, but how it changes the behavior. If that stick doesn't slow the dog down, then no matter how scary it looks to us, it's not functioning as a punisher.

It's the same with reinforcers. If the dog won't take the treat, no matter how delicious it looks to us, then in that situation the treat is not a reinforcer. I see people petting dogs effusively, under the assumption that they are giving the dog pleasure. If you look at the dog, though, it's showing the white of an eye, or ducking away from the pat, or shaking off the unwanted contact. At that moment, with that dog, your effusiveness is definitely not a positive experience.

Technically, a reward or a punisher has no specific definition; it's just anything we've chosen that we think our learner might like or might avoid. Only the individual doing the behavior can truly tell what sort of "postcedent" or subsequent event any particular item might be.

The behaviors of seeking apparent aversives or avoiding apparent rewards illustrate the often-misunderstood subtleties of B.F. Skinner's thinking. His vocabulary deals with processes and outcomes, not with specific items or events.

Find out more

Want to delve deeper? I recommend a new textbook, *Behavior Analysis for Effective Teaching*, by Julie Skinner Vargas. It's the best explanation I've ever read of operant conditioning. It goes back to Skinner's first discoveries. It goes forward to modern training by integrating throughout the text the uses and functions of secondary reinforcers as used in clicker training and TAGteaching (technologies that previous behavior analysis textbooks either get wrong or ignore). The book is aimed at grade-school teachers, but it serves any of us who teach anyone or anything. It is a college textbook— long, complex, expensive—but worth having on your shelf and dipping into often. There's enlightenment on every page.

Attachments

Like humans, animals have preferences for
one member of a species over another. And it doesn't even
have to be a member of the same species for attraction and
companionship to develop. Though science may
not acknowledge these preferences, such
relationships are hugely reinforcing.

February means Valentine's Day, a happy time for me as a child. I lived with the principal of our school, Mrs. Sturley, who set up a card table in the parlor so I could spend hours pasting together the paper, lace, stand-up figures, stick-on hearts, and lovebirds to make 21 special valentines for my 21 classmates. Sometimes I signed mine; sometimes, daringly, I wrote, "Guess who?"

Often I also got to help making my classroom's special valentine box. The box was covered with pink or red paper and silver or white decorations, with a slot on the top through which the addressed, enveloped valentines could be "mailed" during the days preceding February 14.

Toward the end of Valentine's Day, the valentines were dumped out of the box (such an abundance!) onto the teacher's table. Then teacher-appointed postmen passed them around to the rest of us. Theoretically, everyone made a valentine for everyone else so each student received 21. Even so, some children got more, and some got less. So how did the shortages happen? Maybe it was due to the boys? I'm sure not all the boys sat indoors making all the valentines they were supposed to—more likely they made one for the most popular girl or some other favorite, a few joke cards with snappy remarks on them for other boys, and then were out the door.

Girls also had their preferences, and one way to express them was to give valentines only to people you really liked—or, alas, to withhold valentines from people you actively disliked. Those "Guess who?" valentines kept your preferences private, after all.

Animal friends

So what about animal preferences? In *Reaching the Animal Mind*, there's a chapter about the question of preferences and long-standing, individual attachments between animals, discussing cats and dogs, horses, cattle (surprising, that news), and my own research on wild dolphins. Here's the opening of that chapter, as my valentine to you:

> *We are a little presumptuous about individual friendships and preferences among our domestic animals. We assume that because we like each and every animal, they must like each other. A very common complaint of pet owners is that they added a new cat or dog to the household, and friction ensued. These two dogs hate each other. This young cat is pouncing on the old, tired cat with ever-increasing glee. What can the behaviorist or trainer do to stop that? While I sympathize with the issue, I sometimes sympathize more with the pets. Who asked them if they liked this new individual? Perhaps they were never meant to be friends.*
>
> *Of course our domestic animals can indeed form intense attachments, not just with us (as we perennially hope and assume) but with other animals, both within and across species. Animals, like people, have preferences for other individuals that can only partly be explained by reinforcement, and for which we have no particular evolutionary explanation either.*

In about 1985, I acquired my first border terrier, named Skookum (a Northwest Indian word meaning sturdy and useful, but not beautiful). When Skookum was just a puppy, he spent an afternoon playing with a half-grown German shepherd named Orca. A few months later, Orca and her owner visited my house and Skookum and Orca played again. That was it: two encounters. About three years later, I took Skookum to a lecture by a visiting dog trainer. The room was jammed with people and dogs. Skookum, normally respectably behaved in public, suddenly went berserk, pulling on his leash, whining, jumping up and down, trying desperately to get me to take him to something across the room.

"Look, it's Orca, Orca's here!" Indeed it was Orca. Orca was now a big grown-up search and rescue shepherd, looking very different from her younger self. Alas, she had zero time for him now, but Skookum, in spite of their minimal contact, would never forget her."

You can insist on good manners; you can't insist on love

I think we can reinforce friendly behavior among our pets, and thus reduce bickering. I also think we can give them the right to their own preferences. The tired old cat should have an elevated box to retreat to when she doesn't want to be pestered; the busy young cat should have strenuous targeting games to use up her energy. Dogs may be a pack, but they should all have their own separate places to sleep, and all are entitled to some individual attention and individual downtime as well. Where attachment exists, though, we can at least respect it, and give those friends time together, whenever it's possible.

Animals and Grief

Moving, Karen explains, puts us on an extinction curve because we're separated from the reinforcing stimuli we've grown accustomed to, like the window seat where you have your morning coffee. No wonder we get cranky and miserable. That's a loss from a behavioral standpoint. But what are the biological consequences of loss—of an animal's friend, offspring, and so on?

In the aftermath of Hurricane Katrina, we were all moved by the TV scenes of lost or abandoned dogs hanging around their flooded homes, some fearfully evading capture, others swimming desperately after the rescue boats. They had lost their familiar lives. They surely missed their familiar people. And what if there had been more than one animal in a house? Do animals miss other animals? Do animals grieve for each other?

Last year a scientific colleague, Jane Packard, wrote me to ask if, in my opinion, animals that were separated or that lost a close animal companion felt anything similar to human grief. Jane had been looking at biochemical changes in the blood and brain of animals that appeared to be bonded and that then experienced separation and loss for one reason or another. She thought that evidence for grieving was certainly there. What did I think, from a behavioral standpoint?

I looked at the question first from the reinforcement angle (the effects of learning and life experience of the individual) and then from the ethology angle (the effects of innate or genetically based behavior in that species).

We humans don't just miss people who are gone from us. We miss places and things, too. B.F. Skinner wrote a lovely discussion of homesickness. When you move to a new house or a new town, you "miss" the old place, because, in living there, whether it was great or lousy, you developed a set of reinforcers during your day. Making and sipping the coffee. Walking down the street to get the morning paper. Turning on the TV news at the end of the day.

The surrounding stimuli for each particular reinforcer also become discriminative stimuli, which are conditioned reinforcers in themselves. Therefore, when you move to a new place, you lose a very well-established set of reinforcers—the sight, placement, look, even the smell, of the kitchen cupboards, the newsstand, the front door of your apartment or house, and so on. Enduring the extinction of an established way of acquiring a reinforcer is a very aversive experience for all organisms. Whenever you move to a new habitat and lose access to a set of familiar reinforcers, you will feel emotions ranging somewhere from restless irritability to complete misery, at least from time to time.

Even if you are going from prison to a palace, the extinction process for established reinforcers will occur and will have to be endured. I have taken comfort from this insight many times and used it to practical and personal advantage in my peripatetic life.

So, even for animals, leaving a familiar environment is bound to produce symptoms of the extinction process. Losing a species mate or companion may be a wrenching experience if only from the sudden extinction of access to familiar sources of reinforcement.

What about innate behavior? We accept that the biologically determined ties of mare and foal, say, or pairs of bonded birds, are often very strong. We don't feel surprised when a cow bellows for her weaned calf for a day and a night. We tell ourselves that her distress, whatever its nature, is short-lived, and therefore not at all like human grief. (My feeling is that emotion is emotion—handed out to different species in differing amounts perhaps, but causing the same internal sensations while it lasts.)

I think, however, there is a form of attachment in animals that scientists have not yet taken much of a look at: friendship. We can't yet explain these preferences through evolutionary theory or reinforcement history. In Germany some years ago, a herd of cattle, by some rich owner's whim, was allowed to simply exist and reproduce with no individuals being removed. Ethologists studied this herd over a period of years. One discovery was that the cows had particular friends; those friends (as with humans) often dated back to their "school years." When they were all calves together, female calves associated closely with some age mates and avoided others, and those friendships (and antagonisms) persisted throughout life. Isn't that a lot like people? I am a grandmother, but I still have friends whom I first met in the third grade.

Another interesting finding from that cattle herd study concerned the herd sire, the dominant bull. In this undisturbed herd, the lead bull was not purchased from elsewhere as bulls usually are, but rose to his position after growing up in the herd. The investigators noticed that around midday this bull always grazed next to one particular cow. Why? His favorite in the harem? Not at all. Turns out he was in the habit of having lunch with his mother. Those apparently brief attachments between cow and calf might not be quite as brief as we like to suppose.

And of course the depth of the parental attachment can vary, as it can in people. When I was breeding Welsh ponies in Hawaii one of my broodmares, Lyric, had several foals for me, among them a dark bay filly I named Sonnet. I sent Sonnet to California to show and sell, as a two-year-old. She was a sturdy, cobby Welsh—very pretty, I thought. I did not know that the fashion in California was for longer necks and more hock action; no one bought her. I left her there with other breeders for a year or so and then brought her back to Hawaii to join my broodmare band again.

When she was unloaded from the truck, her mother, Lyric, two fields away, saw Sonnet and set up a tremendous uproar of screams and whinnying. The two were reunited in great joy (more joy on Lyric's part than on Sonnet's, I thought). Lyric had seen her other foals come and go with aplomb. Her handsome son Telstar was probably in plain sight, too, another two fields over. Apparently, however, Lyric was strongly attached to this particular daughter.

What about dolphins? In my years as curator and head trainer at Sea Life Park in Hawaii, I watched and often caused many removals of one or another animal from a group; I oversaw introductions and reunions as well. Losses inevitably occurred from illness and death, too. Usually there was no discernable emotional response to the absence. Sometimes the group even seemed to be relieved: "Ding Dong, the witch is dead!" Sometimes, however, when an attachment had formed between two individuals and one died, the remaining individual was truly grief-stricken. At the Seaquarium in Miami, the death of one of two killer whales plunged its tankmate into a terrible state that really endangered its health.

At Sea Life Park we had a pair of Pacific spotted dolphins, Hoku and Kiko, male and female, who performed as a team and were always together for many years. (Read more about Hoku and Kiko in *Lads Before the Wind*.) They certainly liked each other. They swam in synchrony, copulated frequently, and ignored or rejected other animals.

Then Kiko died suddenly of an undiagnosed kidney ailment. Hoku swam for days with both eyes shut, as if he did not want to look on a world without Kiko. Feeling very sorry for Hoku, we gave him a new companion of his own species, a young spotted dolphin female named Lei who had been living in one of the show pools with a bunch of spinner dolphins. Lei immediately joined up with Hoku. Hoku was polite; he allowed her to swim with him, he did not steal her fish or act aggressive in any way, and he even began to look around again. But for a week he kept one eye shut on whichever side Lei was swimming on.

So, if those abandoned and lost Louisiana dogs had canine housemates, do they miss each other? Sometimes not. Sometimes, I think, very much. For more than a few reasons.

Hidden Reinforcers—for Things You Don't Want!

When problems arise with our household pets, so does the question: Who's training whom? If you discover you're the trainee, the cure, Karen says, is in reversing the roles.

For three years, I served on the board of the B.F. Skinner Foundation. The annual board meeting is held in Boston and usually there's a reception afterwards at the Skinner family home in Cambridge, for members of the board and other people who share an interest in the work of the foundation. At the most recent reception, I was talking to a woman I'd met at the same gathering two years earlier, a museum curator with a special interest in scientific instruments—an interest I share. On this particular evening, however, she didn't want to talk about her work; she wanted to talk about her cat.

It seems that when we had first met, two years earlier, this woman had also wanted to talk about her cat, and particularly its habit of jumping on the kitchen counters while she was cooking. She said that I had advised arranging some other, more suitable perch from which the cat could watch what's going on in the kitchen, and then clicker training the cat to use that perch. And she had done that.

Meanwhile, she had acquired a second cat that had proved to be everything a cat lover could ask for. Cat Two never got on the counters. Cat Two, wanting attention, would lie down next to her and purr while she petted it.

Cat One, on the other hand, enjoyed its special perch in the kitchen but still jumped on the kitchen counters. The owner had followed advice (not my advice) to spray it with water, but now, even when she got out the spray bottle and spritzed Cat One, it just crouched, and wouldn't even jump down. The cat had also developed the habit of running around the bedroom knocking things off the bureau after my acquaintance had retired for the night. Not only that, the cat knocked pictures crooked (something no doubt particularly annoying to a museum curator).

The news about the pictures made me laugh, which was not a polite response. I guess I was supposed to solve the problem, but in fact I had no advice. The cat had trained her. In the kitchen she could easily be provoked into playing a game of "Chase the Cat." Now the cat had developed other ways of producing a round or two of this exciting game, even at times when the owner was not normally active, such as at bedtime.

Cats play chase with each other, and in multi-pet households often develop chase games with dogs or even with house rabbits. Nothing unusual about it, except that the human had been suckered into more and more escalations of the game without even knowing she was reinforcing the behavior she didn't want.

See cats training a dog—on video

This story reminded me of Joan Orr's delightful clicker DVD, *Clicker Puppy*. The program shows children, fourth- and fifth-graders mostly, clicker training an assortment of puppies to do all kinds of things. Joan and her husband David and their girls have two cats, one white, one gray— handsome young males adopted as adults from the Humane Society. While shooting the video, Joan brought a four-month-old Portuguese water dog puppy into her house for a week. The puppy, of course, investigated both cats at once. Joan videotaped the initial encounters and the subsequent events.

In their first meeting, the white cat lay down, and responded minimally to the puppy's nosings. You can see this on the DVD. By and by, the puppy stopped trying to get the cat to move, and lay down, too. In short order the white cat and the puppy became "TV friends," often resting together by the sofa in the evening, maybe even cuddling.

The gray cat—and you can see this on the video, too—responded to the first advance by jumping up and swatting the puppy and then running off a little. Hurray! The puppy gave chase. The cat wasn't particularly frightened; from the beginning its movements looked play-related most of the time. In short order both the puppy and the gray cat enjoyed a good game of play-chase and play-fighting when they happened to meet.

The museum curator's two cats have taught her to play their own preferred games. One likes petting, and gets it. One likes pursuit, and gets it. Being scolded and spritzed is not a detriment; maybe it even adds to the excitement (see "Can a Punisher Also Be a Reinforcer?" page 127). And the behavior is so well trained! Two years, at least, of busy paw work have gone into it.

What are you actually reinforcing?

Dog behaviorists like to explain to pet owners that many of the things dogs do that we regard as problems are just natural behaviors in dogs. Yes, and many of the problems pet owners complain of are natural behaviors that have become exacerbated through inadvertent reinforcement by the owners.

Dogs pull on the leash because owners follow where the leash pulls them. From the dog's standpoint, as clicker trainer Carolyn Clark says, "That's how you walk your person." I once saw a woman who trained her Newfoundland to bark menacingly at her every time she went in the kitchen. How? Sooner or later, after telling it "no" and "quiet" and "go lie down" over and over, she gave it something to eat. Then it was quiet and went away temporarily. These are interlocking systems.

I'm not a pet behavior counselor. It's not my job, and in any case I'm not very interested in remedial work; I like building good, new behaviors better than fixing bad, old ones. But I see a big element of hidden reinforcers in many behavior problems—not all, but many—that people bring to behaviorists: running away, barking, nipping, jumping up, pulling, and so on. Maybe any problem should be addressed first, not by investigating the dog, but by investigating the environment. What reinforcer, or what long, variable ratio schedule, is maintaining this behavior so strongly?

Do your pets do something you wish they wouldn't do? Ask yourself, "Am I actually reinforcing it?" And think about what you could be reinforcing, instead. If you have one of these persistent behavior problems at your house, maybe you can solve it on your own.

1. Figure out what is actually reinforcing the behavior. And stop doing that.

2. Establish some other behavior that is a more acceptable way to earn reinforcement.

3. Use your reinforcing behavior to reinforce the alternate behavior.

That's all there is to it.

Isolating the behavior you are doing—that maintains the nuisance behavior—is counterintuitive and so it's difficult. We naturally want the pet's behavior to change, not our own. Emma Parsons's book, *Click to Calm, Healing the Aggressive Dog,* has wonderful tips on noticing what you are doing that reinforces the behavior you don't want and on changing your behavior so that it has a new meaning for the pet. Aggression doesn't need to be your problem to benefit from Parsons's advice on ways to use your responses to build the behavior you like.

Reinforce Every Behavior?

*If mention of variable rates of reinforcement makes
your eyes glaze over, Karen, with the help of a hound dog,
explains it in a way you'll be able to use.*

In 2006 I gave a workshop at the annual meeting of the Association for Pet Dog Trainers (APDT), always both an honor and a pleasure. In the workshop I demonstrated an exercise I'd learned at an earlier APDT meeting from Massachusetts trainer Tibby Chase for teaching inattentive dogs to walk politely at a person's side. The exercise involves targeting and shaping and works even if neither the handler nor the dog know anything about clicker training.

APDT had arranged for a pet owner to bring three friendly but largely untrained dogs. None of the dogs were accustomed to being in public, and while they were fairly quiet, they were, of course, trying to smell everything and greet everyone, pulling on their leashes and paying very little attention to the person holding them. The owner found a volunteer handler for each dog so I could put them through the exercise, one at a time.

Game setup

I set out about 10 circular colored floor markers (the kind that are used in kids' soccer practice) in a straight line about 4 feet (or 2 paces) apart across the front of the room. I asked the handler and dog to start at one end of the line and walk to the other; the only instruction was that every time I clicked, the handler must stop and give the dog a treat.

If the dog was walking at the owner's left side, I clicked just before they came to the next dot. By the fifth dot I didn't have to worry about where the dog was. I had deliberately set the dots so close to each other that the dog hardly had time to be distracted or move away between dots.

At the end of the row I asked the handler to turn around and bring the dog back along the line again. At first the dog's attention wandered during the turn around, and either the dog or handler might be pulling on the leash, but as soon as they started down the line again, the dog fell into position. Click-stop/treat, click-stop/treat, click-stop/treat all the way past ten dots. By now the dog was staying next to the handler on purpose, and the leash was slack between them.

So far, I was using a continuous schedule of reinforcement. The dog was doing what I had in mind, the click was marking it over and over, and the click was always followed by food.

It gets trickier

Before the next pass I stepped in and removed the third, fifth, and seventh dots. I had thus raised one criterion: the distance. Now there were three gaps in the line that were longer than before. "Sometimes, dog, you may have to walk a little further to get to a clickable moment." Just as I expected, the first dog strayed a bit in the new gap. Then as the dog and handler approached the next dot, if the dog was nearer the handler again, I clicked. I was shaping the behavior of "Walk next to the handler for longer and longer distances." Usually by the time the handler and the dog hit the gap between the sixth and eighth dot, the dog was once again glued to the handler's side and remained so from then on. One dog, however, a large hound mix that had been the most inattentive and pull-happy of the three at the beginning, did need three passes down the line to stay at heel during all the longer gaps.

You might say I was still reinforcing the behavior continuously, because I definitely clicked every time the correct behavior occurred: every time dog, person, and dot were in close proximity. But from the dog's standpoint, it was getting reinforced on a predictable basis, and now, suddenly, it was not so predictable. The dog must try a little harder, maintain the behavior a little longer, to find out how to get the click to happen again for sure.

Shaping schedules

During shaping of a new behavior, each time you establish the behavior, the dog is being reinforced on a continuous schedule: it does the behavior and it gets the click/treat. As soon as you want to improve the behavior, however, and you raise a criterion, the dog is on a less predictable schedule. The requirements are a little different, and the behavior probably will not get reinforced every time. From the dog's standpoint, the schedule has become variable. When the dog is meeting the new criterion every time, the reinforcement becomes continuous again.

Marian Breland Bailey told me she called this a "shaping schedule." It's a natural part of the shaping process. Reinforcement may go from predictable to a little unpredictable back to predictable, as you climb, step by step, toward your ultimate goal.

Sometimes a novice animal may find this very disconcerting. If two or three expected reinforcers fail to materialize, the animal may simply give up and quit on you. You can see this clearly on the video of my fish learning to swim through a hoop. When three tries "didn't work," the fish not only quit trying, he had an emotional collapse, lying on the bottom of the tank in visible distress. He offered no more hoop-swimming; scientists would say the behavior had extinguished.

Recovering extinct behaviors

Extinction does not erase a behavior; once learned, it still exists in the animal's nervous system. There are a number of ways to recover a behavior that has gone into extinction, such as reducing your criterion (going back to a dot every two feet), or simply asking for some other well-learned behavior, or waiting an hour or a day and trying again. But perhaps the most graceful way is to build a little confidence, a little resilience in the animal, by introducing a little variability in the reinforcement schedule on purpose, but very tactfully. The animal mostly gets the reinforcement it expected for the behavior it is just learning, but sometimes it has to do the behavior two times, or go twice as far, or twice as long, for a single click. This is the situation I was creating with these naïve dogs by removing an occasional dot: sometimes the dogs had to go the usual distance, and sometimes twice the usual distance.

At first each dog had thought the game was over, but then they discovered it was still working. Both their confidence and the strength of the behavior increased. By the fifth pass down the line, each of the three dogs looked like a polished obedience class graduate: locked into position next to the handler on a nicely loose leash, tail up and waving, head cranked around to look eagerly at the person's face, watching for the next magic moment when a click-stop/treat might occur.

Tremendous progress

Just for fun, when the last dog, that big hound mix, came down the line perfectly, obedience prance, turned head, and all, I grabbed up a few more dots and laid them out several yards apart across an empty section of the ballroom toward the distant entrance. From the end of the line I sent the handler out across the empty spaces, with the dots as targets to guide her. With just two or three clicks and treats each way, the hound walked nicely at heel, gazing up at her eagerly, clear across the ballroom and back. This easygoing dog could now accept quite large increases in criterion, and still give the behavior so well that he was on a continuous schedule again. Good boy!

Variable ratios

So that's a place where a variable ratio of click/treat to offered behavior occurs: when you are selectively reinforcing better or stronger or different behavior. It may happen again when you are adding the cue. Some behaviors may be reinforced and some not; from the animal's standpoint, it is not sure why, and it must be a little resilient about those missed clicks to figure out how to meet the new criterion. Again, when the behavior becomes part of a longer repertoire or rolled into daily life, and natural reinforcers take over, reinforcement may be erratic, and consequently (in my view) on a variable ratio schedule. Yet the behavior is maintained.

Once a simple behavior has been learned, a long and unpredictable schedule can in fact maintain behavior that you *don't* want, with incredible power (see "Hidden Reinforcers," page 138). Inadvertently people train cats to get them up in the night, dogs to pull like freight trains, and children to have tantrums. They do this by holding out for some of the time and then giving in, feeding the cat, going along where the dog wants to go, or buying the candy in the supermarket, on an irregular basis. Casinos, believe me, use the power of the variable ratio schedule to develop behaviors, such as playing slot machines, that are very resistant to extinction despite highly variable and unpredictable reinforcement (see "Jackpots: Hitting It Big," page 105).

So—where do you deliberately use a variable ratio schedule of reinforcement? In raising criteria. For building resistance to extinction during shaping. For extending duration and distance of a behavior (ping-ponging, as Morgan Spector and Corally Burmaster say).

Where do you *not* use it?

Never purely for a maintenance tool. Behaviors that occur in just the same way with the same level of difficulty each time are better maintained by continuous reinforcement, or by reinforcing in various combinations with other behaviors, than by deliberately letting satisfactory behavior go unreinforced.

Never for maintaining chains. I once had the privilege of co-presenting a workshop with Debi Davis and saw her service dog, a papillon, jump down from her lap to pick up and bring back to her a dollar bill she had dropped. Debi promptly clicked and treated, and then told me people routinely remonstrated with her for doing that, saying that the behavior should *not* be reinforced every time. But this was a chained behavior involving multiple steps. The environment provided the cue for each step of the chain. (See money fall, jump down. Reach money, pick it up. Got the money? Take it back to Debi, and so on.) Each cue reinforced the behavior that preceded it. But failing to reinforce the whole chain at the end of it would inevitably lead to pieces of the chain beginning to extinguish down the road. Debi was right. Pay the pup for that great job!

Never for discrimination problems such as scent articles. If you are asking the dog to make a choice between two objects or stimuli, you have to tell him when he's right; putting him on "twofers" just punishes correct answers.

I was very happy with the super performance of my three pullers at APDT and the demonstration of how oscillating between continuous and intermittent reinforcement allows you to raise criteria extremely fast. That is, until a trainer complained afterwards that I had used the wrong dogs. "It would have been a better demonstration," she said, "if they hadn't already been so well trained."

The
Bad
Stuff

Y ou strive to be a positive trainer, but, since punishment is determined by the receiver and is unpredictable in its effects, it's not always easy to tease out what just happened. You thought you were communicating, but your dog is cringing and looking away. Did you just punish your dog?

Karen offers her insights into "aversives" (the daily hassles we all face and try to avoid) and punishment and its insidious tainting effects on the relationship with the trainer. She offers surprising cautions about using a supposed reinforcer (praise) and delivers a heartfelt and astute critique of what's wrong with the "long down" (and what to train instead). For her, training with reinforcement "involves creating a climate of security in which it is safe to learn new things, and safe to rely on what you've already learned." It's the best antidote to punishment.

Shaping a dog to love knocking down piles of objects helps him overcome fear of loud noises in daily life (see p. 163). Photo: Nina Mortensen

Aversive or Punishment?

Aversives are part of life; punishment
need not be. What's the difference?

A list exchange started with the following post from a trainer who didn't advocate training with punishment, but needed clarification about some definitions:

Q: Can you teach everything without punishment? By punishment I mean "correction," which I translated to "punishment" in my question. Maybe that's the problem? I've been thinking that most corrections, even if given gently, are "punishments." But maybe I'm wrong and they are only that, corrections?

In my case, I'm a veterinarian and I must teach a new employee to look at stool samples. I reward her each step of the way, several times, and she seems to understand. I watch her perform each step successfully—until she puts alcohol instead of sugar in the solution. What can I do?

I don't mean to belabor this, but I just want to be sure I'm not punishing the animal (dog, horse, person).

Aversive or punishment?

There's a difference between aversive events and punishment. Life is full of aversive events—it rains, you stub your toe, the train leaves without you. These things happen to all of us, and to our pets, and we don't control when or if they occur. In general, all that we learn from the inevitable aversives in daily life is to avoid them if we can.

On the other hand, a punishment is something aversive that you do on purpose. It may be contingent on a behavior, and it may stop or interrupt that behavior—which reinforces *you* for punishing, so watch out for that.

The effects of punishment

But a punishment does *not* have a predictable effect on the future. We make false assumptions when we declare, "I really taught him a lesson." Research has shown that punishing a behavior may change that behavior in the future, may not change that behavior, and/or may change some other behavior. (Murray Sidman did a lot of this work.)

A punishment might not happen at the same time as the behavior (that's what contingent means). If that happens, the punishment is just an inexplicable aversive, which then may become associated with you more than with any behavior.

Be careful with assumptions

Removing punishment from your tool kit is *not* the same thing as removing all aversives from your learner's life. Many non-clicker trainers leap to that conclusion, point out that it's impossible to remove all aversives—dogs wear leashes and get shut in crates, clicker trainers use the Gentle Leader—and conclude that clicker training is not punishment-free. But that initial assumption is wrong. While all punishment is aversive, not all aversives are punishment.

My dogs spend a lot of time sleeping under my computer, bored. Is that aversive? I suppose that sometimes they'd rather be doing something more exciting, but so what? I have my job, they have theirs, and that's life. I am not punishing them, but they are experiencing the reality, sometimes aversive, of our life together.

Ending a training session

Since clicker training is so much fun, it's true that ending a session is a downer. Sigh! The solution is to end the session so that it doesn't come as a complete and unexpected deprivation, which *could* impact behavior. End with a clear "All done" signal that always means, "That was fun, and now I have to go do something else." I use the Hawaiian word "pau" (finished, done) and a hand gesture, and my dogs accept that cheerfully. If you want to make the end of the session even less painful, after you give your "end-of-session" signal, hand over a toy or some other reinforcer to ease the moment. With dolphins, we scattered three or four extra fish around, a good-bye present. By the time the dolphins had picked up all the fish, they probably felt better.

Ending a fun session is more of a problem with a green, inexperienced animal. If you have thousands of training sessions under your belt, or fins, you probably don't worry, as there will always be another chance.

If your learner is *glad* the session is over, wags, romps, laughs, or plays after the session, but not before or during, you must think about that. I see a lot of performance dogs with that reaction.

Get real

In real life you have to wade in with an aversive to stop something from happening. If you must yank a baby away from the light socket, stop a dog from grabbing the roast chicken off the table, so be it. Animals do reprimand (the official biologist's term) their young and each other. You'll have interrupted or stopped a dangerous event. Just don't kid yourself that you've taught or guaranteed any particular change for the future.

Stepping on the Food

You're training "leave it." You drop a bit of food, the
dog lunges toward it, and you cover it with your foot. Are you
just managing the environment, or is this negative
punishment, taking away something desired?

The implication is that if it's negative punishment, then a good clicker trainer shouldn't need to use it. And that raises another question, "How can you train 'leave it' without stepping on the food?" What an interesting discussion!

Is it or isn't it punishment? A matter of opinion

All dogs experience aversives in their daily lives, such as tripping over a log or having the ball roll under the couch. Do the dogs learn from these episodes? Not so you'd notice. So, your stepping on the food may just be one of life's aversives for your dog, as if the food had fallen through a crack in the floorboards. Oops. Oh well.

All punishers are aversives, but not all aversives are punishers. However, punishment, like beauty, is in the eye of the beholder. Was it a punisher? It depends. There are two tests. The behaviorist's test is, "Did the behavior subsequently become less frequent or disappear?" You may not know the answer without numerous repetitions and, as with many punishers, you may not get the outcome you predicted. After a few experiences, instead of giving up lunging, the dog might lunge even faster, to try to beat your foot to the food. If the lunging is increasing, then moving your foot has become the cue to lunge. The behavior is reinforced if the dog beats your foot, so he's on a variable-ratio schedule, which maintains the behavior.

The intent, however, is to diminish the behavior. In this case, the behavior at first might have been "Look at the food." Maybe smelling or digging at your foot. The lunge could be a completely new behavior, shaped by consequences: sometimes the dog got the food, at least at first.

The ethologist's test to see if stepping on the food is an aversive or a punisher in the eyes of the dog is the behavior of the dog. "Did the dog cringe or draw back, or, if the dog just hesitated, was the facial expression one of anxiety?" If so, then you know the dog probably experienced a punisher.

Shaping with variable-ratio schedules

Is experiencing punishment an inevitable part of the shaping experience? I don't think so. In shaping, you need to put the organism on a variable-ratio schedule so that you can reinforce selectively, choosing one behavior over another. You must build at least a little resistance to extinction, because you don't want the animal to quit every time you fail to reinforce a behavior just once or twice. (That is what was meant by "twofers" way back in the early days—a step in creating tolerance for a shaping schedule. Somehow it got exploded into a fixed-ratio schedule as some kind of absolutely necessary maintenance tool, which, of course, it is not.)

In an organism's very first experience of shaping, even a very small increase in the ratio—a single experience of going from 1:1 to 1:2—can put a naïve learner into an extinction curve. The learner finds it very punishing, too. Think of the fish video where my cichlid faints the first time he fails to get a treat for swimming through a hoop—when he was sure he would succeed. I've had the same experience with a naïve dolphin.

By the time we start training "leave it," the dog is probably way past being horrified at having to try something twice to get paid once; a dog's life is full of variable outcomes and dogs become resilient pretty quickly. So, most of us do our shaping by playing within the bounds of a low-ratio, variable schedule. We raise criteria slowly enough so that the learner has a good chance of success most of the time.

On Virginia Broitman's DVD *The Shape of Bow Wow*, there's an interesting sequence on shaping a papillon to stand on a box. Both the trainer and the dog are skilled at the game. The dog tries something new, gets clicked for it a few times, and then Virginia raises the criteria. Oops, the next try doesn't work, so the dog intensifies or varies the behavior. Yay! Success again for the next few tries, and so on. It's not an arbitrary schedule, of course. Virginia is not counting her clicks. She reinforces everything that meets her criteria, raises her criteria quickly, and the dog wins a good bit of the time, but the behavior keeps changing in the direction of the ultimate goal.

This dog is *not* on an extinction curve. Virginia is *not* letting things get to that point. Indeed, unreinforced behavior drops out and reinforced behavior increases, but with the dog's conscious participation. There's no punishment involved; the dog is having fun.

So, it is possible to shape behavior within the context of variable reinforcement schedules that are tolerable without shifting into deprivation and punishment or putting the animal into an extinction curve. The Baileys simply call this a "shaping schedule."

Tolerance varies—long reinforcement schedules versus extinction and stress

You can build into your dog or other learner a tolerance for long reinforcement schedules, where it takes many tries to find out how to get a click. Some years ago I was in England, giving a seminar for John Fisher's associates. I asked if someone wanted to demonstrate the box game on stage. A woman came up with her nice Border collie. The audience chose the behavior—going under a chair. After getting clicked once for interacting with the chair, the dog tried a dozen or more different behaviors related to the chair, to the point where the audience was crying out, "Stop it, the poor dog, that's enough." But the dog seemed calm, and the owner knew its capacity; the decision was hers. Next, the dog poked its nose under the chair, got a click, peered under the chair, click, understood the behavior, skipped its treat, scooted under the chair, and came out with all flags flying! The ratio was challenging, but not too hard for that experienced dog. Every single thing it tried that did not get clicked, it abandoned at once. What we saw was *not* extinction, but a decision, a shift in the search pattern, if you like. The actual behavior—looking for that click—stayed very strong.

You can, of course, make the schedule so demanding that your animal does indeed show stress, even though it is making progress. I have read comments by non-trainers saying that they see people shaping with the dog in a state of real anxiety, lips drawn back, tongue out, stress lines in the cheeks, and so on. The simple explanation would be that the trainer has raised the criteria much too fast, and the animal is, indeed, on an extinction curve, but I don't know. I've never actually witnessed this myself. When I first read an article by another trainer saying that his shaping consisted of "surfing the extinction bursts," I was shocked. Yes, you could shape that way, but, ouch, you'd need to build a very hardened dog. It seemed sad to me that someone had figured out a way to weave punishment into the exciting experience of shaping.

A plan to follow

So how do you draw the line? Let's go back to "leave it." If the animal is lunging for dropped food, then you have raised criteria too fast. Start with dropping boring objects, those that the dog would investigate only to see what they are. Poker chips and dead leaves are good examples. Here you can easily get a smiling "Oh, okay" drawback from the object in a few clicks. (You shape the drawback before you start saying "leave it" of course. Click for looking at it but not moving toward it, for example.) Proceed to slightly interesting objects, and then to toys. Vary your behavior, the location, time of day, indoor/outdoor, and so on, build a good reinforcement history for that cue, and then go to boring food (lettuce, carrots), ending up with a recall through scattered bits of steak or whatever your ultimate test item is. At what point do you introduce the cue? I don't know. Maybe when the response is well established with the first objects; that's a shaping decision.

The constructive click

Earning a click is always more powerful than just scarfing up food. I'm sure many of you do the same demo I have often done at ClickerExpo with excitable shelter dogs—clicking the dog for walking at my side, or some other behavior, at a very high rate at first, and tossing the food on the floor. In five minutes, the dog may be totally focused on making me click, and on the treats that come from the click, though it is wading through dropped bits of hotdogs it missed, without a thought of "Hoovering," or vacuuming, the carpet.

So, if you have to manage the dropped food by stepping on it, it *is* just that, a temporary management tool. If you are doing it repeatedly, to prevent the dog from doing the wrong behavior, you are introducing correction into your process. The animal may view it as an accidental mishap (in which case the behavior may persist), or as information: "Oh, leave it, right. I remember now." Or it may be viewed as punishment; the animal's demeanor will help you decide. Punishment has fallout. One kind of fallout is that if you have punishment in your toolkit, you are much more likely to reach for it again, Kay Laurence warns. Punishment is for stopping behavior, not for building it. Be constructive instead.

Hidden Aversives: Are You Punishing Unconsciously?

What a "positive" trainer views as communication, an animal may view as punishment, with long-lasting loss of trust in the trainer and enthusiasm for the training process.

It's a new year. Time for good resolutions, right? Let's resolve to stop punishing our dogs by accident.

"But I'm a positive trainer, I don't punish!" you say. Well, I think we sometimes do, and don't realize it.

I'm not talking about reprimand. That's a social act. A puppy gets too rough, and Mama dog growls. A canine nose gets a bit too near the cheese and crackers on the coffee table, and you speak warningly. That's communication, and it's legal, in my opinion (unless it becomes *all* you do and replaces actual teaching—and I do know pet owners and parents who lean that way).

I'm also not talking about naturally occurring aversives (an aversive is any event that one might find unpleasant). As British clicker expert Kay Laurence has said, "Dogs get plenty of 'aversives' in everyday life. My Gordon Setters regularly run into the door in misjudging the gap, stub their toes, fall off the bed, go to the wrong side of a tree, etc. Hey, that's life. But it's not good teaching."

Jumping up

What I am talking about is aversives that even "positive" trainers use deliberately, to affect behavior. Here's a popular one. A trainer asked me,

"Karen, how about when the dog jumps up? Are you saying it is not proper clicker training (+R) to turn your back (take away attention) and then click him for four on the floor? This is an aversive, right? I only want to be a 100% + trainer, and I use this aversive technique. Please elaborate."

I have a video clip of a trainer doing just that. She's working with a big, yellow shelter dog that jumps up a lot. Twice, the dog offers a sit, and she clicks and treats. The third time, the dog sits, but the trainer waits a bit longer, and the dog jumps up on her. She folds her arms and turns her back. As she does that, the dog cringes back toward the floor, as if it had been struck.

Was that "punishment"? To the trainer, no; she just briefly removed her attention, so what's so bad about that? To the cowering dog, yes, that really hurt.

Punishers, like reinforcers, are defined by the receiver, not the giver. I am sure the trainer thought she was just communicating, not punishing. But in a dog's world, licking faces is a puppy's way of making adults be nice to them, so dogs jump to get near our faces. Turning your back, therefore, is negative punishment: removing something the dog very much wants. The dog's cringing tells us, from the dog's point of view, this was painful.

Now this delightful clicker training experience, in which the dog *was* learning how to be successful, has become a mix of good and bad. Learning, even with food involved, is no longer such a good thing. The trainer herself is no longer an unmixed blessing, either. That's what the introduction of punishment does, whether you want it to or not: it makes the learner insecure. Jesús Rosales-Ruiz and his students have shown us new research from the University of North Texas that suggests it doesn't take much punishment, and it doesn't take long, for that unpredictable mix of good and bad to taint the dog's enthusiasm for attending to you, and for learning in general, for a very, very long time.

"Guiding" with the leash

Here's another popular punishment that I see at ClickerExpo all the time. The dog is on leash, quietly standing or sitting next to its owner. The owner gets up and starts to walk away. These owners typically don't look at the dog, they don't speak to the dog, they just start off, and as they do so, they jerk the leash. That's how they tell the dog, "Come with me." Even if the dog is already moving with them, bang! That little leash pop. Why?

In 2005 in Orlando, I finally got annoyed enough with this to create a ClickerExpo demonstration around it. I picked someone with a large, pleasant young dog, who was doing this leash jerk faithfully every time she sat down or stood up or moved. We got a chair for her to get in and out of. I told her when to move, and I clicked her for giving the dog a verbal cue before moving. I tried to foil that well-honed habitual yank by putting the leash in her "wrong" hand. It was hard for my demo dog owner; she really thought she had to give that leash pop. It was easy for the dog. When I finally stopped the popping by taking the leash off altogether, the dog, almost prancing, eagerly glued itself to the owner's side while she stood, sat, stood, and walked. "Whew! What a relief," said the dog's happy face. Everyone laughed. Obviously, the dog had learned very well how to pace itself to the owner's moves, and was glad to do so. The yanks were completely gratuitous punishment.

Teaching aversives as requirements

Someone, I could see, had put a good deal of effort into making that owner a consistent punisher. "Probably you've been scolded for *not* popping the leash every time," I ventured, and indeed heads nodded all over the room.

So, aversives are used on the owners, to make them use aversives on the dog. And does it do any harm? Oh yes, indeed. In the course of one morning at ClickerExpo, these well-behaved dogs might receive dozens of unexpected aversives. Pop, pop, pop. They get a yank while lying down quietly (surely a good thing to be doing), or while standing still next to their owners (another good thing), or while walking with them as they change direction, or stop at the elevator door, or start down a hall. There is nothing they can do to avoid the pop; even skillfully anticipating their owners' every move does not bring relief. No wonder they wear facial expressions of patient resignation, like most horses.

Major vs. minor?

The question is *not* how major or minor the aversive is. The question is, why use it when you don't have to? In *Learning About Dogs,* Kay Laurence writes: "If I am teaching a dog, I avoid every atom of punishment or removal of something good to get the behavior. It is not a question of how aversive, it is the thought that aversive is a method to get a behavior. The actions are an indication of the thought process that aversives are part of the teaching process. I will say, 'Let's just find another way.'"

We all think we're "positive" trainers. But training with reinforcement involves more than just being nice and more than using reinforcers. It involves creating a climate of security in which it is safe to learn new things and safe to rely on what you've already learned. In this climate, an animal can learn to control itself, rather than being controlled by you. In this climate, rather than just reacting to the environment like an untutored shelter dog, barking at every noise, plunging toward every attraction, jumping on everyone and everything, mouthing and smelling and grabbing, an animal becomes confident and calm. In this climate, having confidence that your cues are meaningful and will lead toward pleasant goals, the dog is trusting and—this is very unscientific—the dog is happy.

A Scaredy-cat Dog

Karen takes to heart the lessons from
"Hidden Aversives" and practices not only banishing all
unintentional punishment but introducing "daily merriment"
into her fearful poodle's life.

I've had a lot of dogs in my life, Labradors and poodles and Great Danes and Border terriers, a Weimaraner, a collie, a golden, and a great mutt named Goulash. But I never had a fearful dog—until the current Dog-in-Residence, Misha, my German spotted poodle. He's 10 years old now, and still panics at every sudden sound or strange sight. If a new person or a new piece of furniture comes into the house, Misha has to lift his leg somewhere. If I'm too preoccupied by work, or, God forbid, annoyed with Misha, he vomits and has diarrhea, usually on the dining room rug, which of course makes me more annoyed.

Help from England

Then I read Kay Laurence's perceptive and amusing book, *Teaching with Reinforcement*. Using her own cluster of sheep dogs and Gordon setters, Kay demonstrates the ways in which our everyday life with our dogs is riddled with both reinforcement and punishment contingencies we may not notice or understand.

I watched Misha for a few days and discovered that a *lot* of my "normal" behavior is punishing to Misha. Even his name can be aversive: "Oh Misha, did you do that *again*?" Misha is entirely manageable, he knows a lot of tricks, his human and dog manners are splendid—but we need some higher thinking here. Goal: no more anxious looks and slinking around the house. I want daily merriment.

I started by seeing if I could help Misha have a whole day without feeling fear. It turns out that I can. When he pulls back and freezes at something—my briefcase in a new place in the car for instance—I used to think "Oh Misha, get over it." Then I'd lift him into the car manually. Now I'm learning to give him time to study the situation and cope with it at his own speed. So that reduces some of our chronic conflict.

Help from Sweden

Next I read *Agility Right from the Start,* by Swedish trainers Eva Bertilsson and Emelie Johnson Vegh. Here are step-by-step instructions for teaching a dog how to play with toys (teaching an owner to play, too, a lesson I sorely needed). I broke out the clicker and some new, tiny, star-shaped treats that Misha loves. I carefully shaped taking and holding a small ball.

We did that for two weeks, ten minutes a night. Then suddenly, one day at the office, Misha found a tennis ball and exploded into ball play, chasing the ball up and down the halls, throwing the ball on his own, dropping it at my feet for me to throw. Yay Misha! Now he plays with all kinds of toys, even new, strange toys that he used to avoid.

Misha and Karen's excellent Christmas present

But wait, there's more. In their book, Eva and Emelie have given us a precise set of clicker exercises for gradually teaching a dog to enjoy making things clatter and bang—knocking down a pile of metal bowls, say. When a dog learns to earn clicks and really delicious treats by making noise himself, then noisy events and mishaps on the agility field—jumps falling over, the teeter banging, loudspeakers nearby, and so on—become far less disturbing. Of course it would be the same in everyday life, too. So that's my

Christmas present to Misha: we're going to do all the exercises in Eva and Emelie's new agility book for teaching a dog to love a commotion.

And you know what? Being obliged to have some clicker fun with my now-elderly dog is Misha's Christmas present to me.

The Perils of Praise

We've all been told to praise our dogs, but is it always what they
want? What is praise good for? What are its limitations?

Early in the summer of 2004, I visited Honey Loring's Camp Gone to the Dogs, in Vermont. Classes were held all day long for various kinds of dog sports and activities. The throngs of vacationers were showing off their pets and having fun experimenting with new activities. All the instructors were positively inclined. Many people carried treats, but clickers were not much in evidence. I saw lot of leash guidance, food luring, and praise.

Does praise reward new learning?

It made me think about praise. There's a long-standing tradition that gushy praise is something dogs like. That praise helps them learn. That praise tells them you are pleased, and that they like that. But is that all true?

Here's what praise can't do: it can't tell the dog exactly what it did right, the way the click does. It comes after the fact, and it goes on too long, so praise *never* serves as real-time feedback during a learning situation.

While praise can't inform specifically, however, perhaps it can inform in a general way, signifying that the whole effort was well done. That's the kind of praise/reward people hand out lavishly to children: a big "Good job!" for anything from washing their hands to getting into Harvard. I once listened to a mother abundantly praising her nine-year-old daughter for having read a whole book. I must have raised an eyebrow, because the little girl turned to me and said, "I get praised for reading. I get praised for being nice to my little sister, too." I nodded. She added, "Of course I'd read, anyway." For that child, praise was a known manipulator, and she had already learned to manipulate back: being "nice" to her little sister—in Mom's presence, at least—kept Mom happy, didn't it?

Children can usually tell when praise is intentional, artificial, and designed with an end point in mind—when it's fake—and when it is spontaneous and genuine. Do you think dogs can't? Is your praise rewarding? Are you sure your dog *likes* your praise? I saw vacationers praising their dogs repeatedly without noticing what the dog was telling them about it in return. I saw dogs ignoring the owner. Looking away. Yawning (a sign of stress). Responding to the torrent of words by finding a fascinating smell in the grass or a terrible itch that had to be scratched that instant: a dog's ways of saying "Enough already! Calm down, you're bothering me."

And praise combined with petting could be even worse. I saw some owners "rewarding" a dog for, say, performing an agility obstacle, or taking a first swim, with praise plus lavish petting on the head and shoulders, even thumps on the back and ribs. In many cases, the dog ducked the advancing hand, and shook itself all over as soon as the hand went away, as if to get rid of the petting sensation. Some dogs looked anxious, or even downright upset. Maybe this did no harm—dogs graciously put up with a lot of our annoying activities—but did it reinforce, or strengthen, the new achievement? I don't think so. What if praise means "Oh boy, they're not mad at me?" I wonder if the custom of lavish praise became ingrained as a training method in the days when most training consisted of correction, correction, correction. When the dog *did* do something right, and you praised, at least the dog knew that for once it had escaped the yank and the yell. Yippee.

When dishing out praise becomes a signal that punishment is not going to happen, at least for the moment, a dog may learn to respond with a glad face, or even with leaps, bounds, licking, tail wags, and eye contact. It looks great, all this licking and bouncing, and we feel that it shows the dog loves us. But from my standpoint, this also looks a lot like what biologists call an appeasement display. "Aren't I cute? Don't hurt me, I'm only a puppy and I just *love* you." (That's one reason dogs, even big grown dogs, like to leap and lick and frolic around strangers in the doorway. From their standpoint, they're appeasing a potential adversary or threat.) So is praise *ever* a good thing? Of course. And I saw a beautiful example during my visit to the dog vacation camp. In a luncheon talk I'd been asked to give, I'd raised the issue: maybe praise is not always something dogs appreciate; maybe praise is by no means a sure way of strengthening a behavior or instilling a skill.

One man was particularly stricken by this idea, and spoke out. Was I suggesting not praising your dog anymore? A terrible thought to him. I understood and sympathized with his distress, but there was no chance for more discussion: camps have schedules, and it was time for everyone to go on to the next activity. He left perplexed, and I too left with a sense of unfinished business.

That afternoon I was taken on a tour of the ongoing activities: a swimming hole; a wonderful "Come when called" class taught by Leslie Nelson; a wildly varied group of dogs exploring agility. (I fell in love with a Cavalier King Charles spaniel sitting next to me in the grass—talk about praise! This dog was a professional flatterer, and I found it very reinforcing.)

Then my guide and I stopped under a tent to watch a class in conformation handling. I recognized the instructor, an experienced judge, from my luncheon audience. I recognized, too, one of the handlers: the man who had defended praise.

He was stacking and gaiting a handsome giant schnauzer. He wasn't coaxing and verbalizing and gushing with praise; he was just giving signals, getting what he wanted, listening to the instructor, turning back to the dog, and going on to the next task. Patiently and attentively, the dog listened to his owner's quiet verbal cues. He moved his feet and held his head as asked, stood motionless as asked. He stayed attentive and on duty while the judge was working with other dogs and handlers. For at least 15 minutes the man and the dog worked together, now on the move, now stationary and stacked, to carry out the teacher's requests. Then the lesson was over. As they hurried out of the ring together, the owner was beaming and waving his arms. I heard a murmur of enthusiastic words: "Thanks for a great job, buddy!" The dog gamboled a little, looking up at him with a laughing face—"No problem, boss, any time!"

And that's when praise is just right. Not when it's a training device, used in the hope of making something happen better, but when it's in genuine thanks for a performance of already-learned skills. Then it becomes a meaningful social exchange, reinforcing to the giver and the receiver both. Great job, guys.

Debunking Dominance Theory

Does the dominance theory hold any water?

Throughout the pet business right now, "dominance theory" is a popular explanation for absolutely anything that happens, from a puppy tugging on your trouser leg to birds flying up instead of down. Conquering "dominance" has become justification for absolutely any punishment people can think up, from shocking dogs to stuffing parrots into the toilet. (Yes, seriously.) And the awful thing is that otherwise sensible people believe this nonsense. Apparently the idea that some animal is trying to "dominate" *you* really resonates. Yikes—gotta stop that, right?

You may be pleased to learn that some British scientists have blown a hole in the whole dog dominance business. Researchers in companion animal behavior in the University of Bristol veterinary department studied a group of dogs at a re-homing center and also reanalyzed existing studies on feral dogs. Their conclusion: individual relationships between dogs are learned through experience rather than motivated by a desire to assert "dominance."

According to these specialists in companion animal behavior, training approaches aimed at "dominance reduction" vary from worthless to downright dangerous. Making dogs go through doors or eat their dinners after you, not before, will not shape the dogs' overall view of the relationship but will only teach them what to expect in those situations. In other words, that stuff is silly, but harmless.

"Much worse," states a nice summary of the research in *Science Daily*, "techniques such as pinning the dog to the floor, grabbing the jowls, or blasting hooters [noise makers] at dogs, will make dogs anxious, often about their owner, and potentially lead to an escalation of aggression."

Consequences

Veterinarians and shelters are seeing the results of this misapplied dominance theory. As one veterinary behaviorist put it to me at a recent scientific meeting, "A puppy has to submit to whatever the owner does; it has no choice. Then around the age of two comes just one Alpha roll too many, and the dog defends itself at last and tries to take the owner's face off." So now the dog is in the shelter. And these dogs are fearful, unpredictable, and very hard to rehabilitate.

Teaching people the power of clicker training is the benign and much more effective alternative. I'm so glad you all are out there, showing people through your own example and your happy, cooperative, attentive clicker dogs that there is a better way.

Why I Hate the Long Down

Karen challenges the rationale for teaching a hallmark of
traditional obedience and offers a kinder and more sensible way
to accomplish the goal—for both dog and trainer.

When I first took a dog to obedience class, back in the Pleistocene, we were given six weeks to teach our dog to obey five basic commands: sit, down, stay, heel, and come. The behaviors were a given; these are the things any well-trained, obedient dog should be able to do. The important thing was not just doing the behavior, but Obeying the Command No Matter What.

The long down

Perhaps the most important of all the behaviors, and the most difficult, was the long down. Could your dog lie down and stay down while you walked away? Could he stay there until you came back and told him he could get up? What if his mortal enemy was next to him, or the instructor walked behind him, or some other dog got up and came over to him? Never mind! He'd better not budge! If he moved, we screamed, "No! Down!" and rushed back and jerked him into position again.

We were encouraged to "proof" our dogs by making them do long sits and downs in distracting areas. People sometimes took this to extremes. Once at a dog show I saw a woman "down" her dachshund two feet from the entrance to a building, with crowds of people going in and out. The owner, her back to the dog, was chatting with friends down the path a few yards, when a man walked past the dog, not seeing him at all, and almost stepped on his paw. The little dog, still in the down position, shrank back a few inches. "Your dog moved," someone said. "No! Down!" she screamed, running back to him and jerking him forward to the pavement's edge again.

Other parts of the old training are fading, but somehow the custom of teaching sit-stay and down-stay for longer and longer durations and in the face of bigger and bigger distractions remains. And I hate it.

A time and a place to settle

I have no objection to teaching dogs to wait patiently when there's nothing else going on—the "settle" exercise. One common and sensible way to do this is to provide the dog with a physical cue, a blanket or mat, that, when spread, means "Relax here; we're going to do nothing for a while." A "while" can be anything from a few minutes to the length of a ClickerExpo lecture; dogs accept the concept easily. Nobody is making a test out of it; nobody is saying "and you have to stay in the same position without moving, and you have to stay motionless No Matter What." The settle exercise is a way of calming the dog down; it is *not* a tool for proofing against distractibility.

Alternatives to the long down

Why not teach the skill of resistance to distractions with a moving exercise, such as walking beside you, or coming when called? For one thing, it's much easier for a dog to learn what he got clicked for when he's moving; it's as if his muscles know what was happening at the time. If he's sitting or lying down motionless and hears a click, how can he tell what it's for? The number of seconds or minutes he's been there? How can he judge that? The fact that near him someone was flapping a towel and he did nothing? Doing nothing is a difficult behavior to distinguish.

Secondly, I think the long down as it is generally practiced is an artificially difficult exercise in itself, because the dog is so vulnerable. The dog has been asked to lie down in an exposed situation, maybe outdoors. The owner is farther and farther away. If it's in a training class, there are other strange dogs around. The dog has no permission to flee or defend itself if things do go wrong, and no one knows better than the dog that things could go wrong.

Look at the facial expressions of dogs learning the down-stay in classes, and especially those being put in down-stays in hallways and parking lots and other public spaces. They are not just a little stressed; they're really worried. And they should be! Now on top of that we add deliberate disturbances, such as oblivious strange men with big feet, to teach them to resist "distractions?"

Distraction resistance

In the video/DVD *BowWow Take II* Virginia Broitman and Sheri Lippman show some excellent examples of teaching distraction resistance with a moving dog. Put out a bowl, upside down, with food under it. Then ask your dog to walk at your side past the bowl without looking at it. Can he do it? Click! Treat! How about a little closer? Could you heel him between two bowls and retain his attention? How about putting him in a sit and calling him to you through a maze of bowls? What if it were just one bowl but it was right side up and you could see the food? Start with a target stick and target him past the temptation, if necessary. The dog is getting clicked and treated for continuing to walk forward instead of turning toward the distraction. That, he can understand.

You can always increase distractions. What if one of the temptations was a cat in a carrier? Or a rolling ball, first in the distance, then right across the dog's path? I think experiencing earning clicks and treats in the face of such temptations gives both the dog and the owner a clear sense of what it means to control oneself. It also seems to me that this kind of training or proofing for ignoring temptations and distractions might carry over more easily to the real world, for example when a squirrel runs by, if it were done in a moving exercise than in a still exercise.

Then, when the dog is already an ace at ignoring distractions, you could ask him to lie on his mat even at the vet's office, or to do a sit-stay while tied to a parking meter while you mail a letter. Now, downing amidst distractions should not be quite so terrifying. And if the dog is a little anxious, here's a comforting touch I've seen used in Europe: fold up his leash and ask him to lie down with his paws on the leash. Now he's not just in limbo, doing nothing, for an unknown length of time. He's Doing a Behavior he can get clicked for: touching a target until he hears a click. Besides, it's got to help his peace of mind. He *knows* you won't forget to come back for the leash!

6

A Confusion of Cues

When you first cross over from traditional to clicker training, it's easy to look back and think of a command as a cue with a threat. It has compulsion: do it or else. But then you learn they're two entirely different things. Cue terminology is intimidating; it's a tangled web. A manila envelope can be a cue? But isn't a cue a word? A cue can also be a whistle, a target, even another behavior. And then there are all the things you can do to a cue: You can switch it, fade it, chain it, poison it (Oops. You don't want to do that). It makes your head spin.

In this chapter, Karen deftly guides you through the world of cues and demystifies what they are and how they work.

Michele Pouliot's springer Deja Vu does her signature "skip" heeling during their "Cleaning Lady routine" to Michael Jackson's "The Way You Make Me Feel" (see page 206). Photo: Brenda Cutting

Bossed Around by My Teakettle

We find comfort and reinforcement in using familiar
objects. We rely on them—until they break. What we don't
realize—and what Karen illustrates beautifully—is what
they're doing to our brains.

Years ago Kay Laurence was visiting Boston and staying at my apartment. Kay and I both like a cup of tea in the morning. I boiled water in an old stainless steel Revere Ware kettle I inherited from my stepmother about a million years ago. Kay made me a present of an electric teakettle. The electric teakettle looked a lot like my old stainless steel teakettle and was the same size, but it had a built-in plastic base that plugged into the wall. It boiled much faster and was a whole lot easier to clean. I threw the ancient Revere Ware kettle away.

The power of conditioned stimuli: danger ahead

We used the new kettle during Kay's visit, and I went on using it very happily for a month or so. Then one sleepy morning I filled the new kettle with water, but instead of attaching the cord and plugging it in, I absent-mindedly put the kettle on the stove, turned on the flame, and...oops. The plastic base immediately made flames halfway to the ceiling. I doused it in the sink, but, of course, the kettle was done for.

I had now learned, however, that an electric kettle was much more convenient—Kay was right about that—so I went out and got another just like it. Another sleepy morning, I again put the kettle on the stove and again turned on the flame. Again, oops.

A more-than-subtle change

I thought there must be a way to own an electric kettle without triggering the responses built into my brain by the shape, size, and color of the kettle I'd been using for decades. I went to the appliance store and looked for something different. Yes! They had electric teakettles. Some of them were shaped like "real" teakettles. But others were shaped like coffee pots. And the coffee pot kinds were mostly vertical—like a jug, and black, not stainless steel.

I bought a nice, black electric jug. The alteration of the conditioned stimuli of shape and color did indeed prove sufficient to protect the new pot from being placed, and burned, on the stove.

I liked that pot. I used it every day. It had its drawbacks, of course; what appliance doesn't? Toasters are worse—at least this pot boiled water while no modern toaster actually makes good toast.

One nuisance I found with the electric kettle was that to pour the hot water out you had to hold down a button, and right next to that button was a hot spot in the pot's plastic side, invisible but hot enough to hurt.

Another irritation was that the kettle took quite some time to heat the water. Also, there was no way to tell when it was done except to look to see if the red light was out. And, it made a great tangle of cord on the counter, which was messy and confusing because of the tangle of cord from the little chopping thing that also lives on the kitchen counter. It was hard to put the pot back on the plastic base; it only went a certain way, so you had to fiddle with it. And you couldn't tell if the kettle was full or empty except by lifting it.

Here we go again

Early this fall, the electric kettle started being unreliable about turning on at all. You had to kind of slam it in a certain way. I decided to replace it. I went to Target and looked at a jillion coffee makers and a few simple jugs. There were none, I noted, that looked like traditional kettles. Perhaps I wasn't the only person who had been conditioned to burn up electric kettles.

I bought an Oster, which looked like a silver coffee pot, and was marked down from $35 to $25, always a good thing. I took it out of its box, washed it, filled it with water, plugged it in, and made tea. Wow! Unlike the toasters, which get even more expensive and even worse at making toast, this kettle actually shows improvement over the previous black jug. The wire coils up inside the bottom so it doesn't lie all over the counter. It is amazingly fast to boil. Really fast. There's a water depth indicator, and a heat indicator, and a friendly ping to tell you when it's done boiling. The pot connects to the center of the base, and it swivels with ease so it goes on its base just right no matter how carelessly you set it down.

I think I'll have a cup of tea right now.

Then I started looking thoughtfully at the microwave oven. "For maybe ten years I have put up with *you*." The inscrutable guide book. The weight/time knob that doesn't work anymore. The 10-step directions for setting the clock. You have to change the clock any time the power goes out, plus twice a year for daylight savings time. But the directions are totally counterintuitive, and I can never remember how to do it, so I have to keep the instructions on the fridge door where I feel a twinge of annoyance every time I notice them. Above all, I hate the irritating "I'm done" beep that will repeat every five minutes *forever* if you don't come and turn it off.

I am tired of being bossed around by this oven. It's like riding a surly horse. Yes, you get some use out of it, but it's a struggle all the way. Maybe Oster makes microwave ovens, too…

A Canary Who Cues

This caged bird sings—and, it turns out, orders his
universe (or, at least, his pizza).

Let me tell you about my canary. I was working on a new book. I was lonely, plugging away in my home office day after day after day. It occurred to me that a singing bird might be company, so I went to my local pet shop to see Loretta, who keeps a lot of birds. I asked her if she could order me a roller canary. German breeders in the last century developed two kinds of singers: rollers, who sing a soft, burbly song like a thrush; and choppers, who have a loud, percussive, brassy sort of song. Loretta said she'd look into it, and I forgot about it.

Six months later I was in the pet shop and a wonderful song lifted itself above the general hubbub in the bird department. I went to see. It was one of the male canaries, whose color was so intense it was like a sign, "I am *yellow!*" and whose voice was similarly attention-getting, "I am *singing!*" I went closer. The cage was on a high shelf, and I was level with the bottom. That canary stopped singing, came down to the floor of the cage, hopped over to my corner so his beak was about three inches from my nose, made eye contact with me, opened his beak, and poured out music. Okay, okay. I get it, I'll take you home.

With the cage and other costs this became a rather expensive proposition, but the bird came home and was established in my office. I soon found out that this bird can sing more or less constantly for approximately four hours in the morning, with other bouts during the rest of the day. (On a good day he can put in an extra hour at night, even in mid-winter, singing duets with the television.) And he's not a roller, he's a chopper. I got the brasses, not the winds; thinking of oboes, I brought home a trumpet.

It was, in fact, a Bit Much while writing, so I moved him to the front of the house where there's more going on and where he seems busy and happy. Like many canaries, he sings when the water is running, when the vacuum is running, and whenever he hears conversation. His songs are interesting, full of variety, beautiful—but really, really loud. I'm used to it, but some people, including company president Aaron, find it appallingly distracting. I usually take my business phone calls in the kitchen to reduce the bird factor.

Clever cuer

I think this bird was hand-raised, because he seems imprinted on people. My upstairs neighbor, Chris Dowd, takes care of him while I'm away. Chris is a very observant animal person. She found out how to elicit wing-fluttering, the food-begging posture of a chick, by presenting her hand to him in a certain way, which for me confirms the likelihood that he is imprinted. Chris pointed out to me that the bird starts to sing when my car pulls in the driveway. Isn't that nice? Once she commented on it, I noticed it, too.

In birds, early conditioning to humans can wreck future breeding attempts (fine with me, as I have no interest in having *more* canaries), but it makes for a good pet. So did I teach him to sit on my hand, and to fly around the room following a laser target? Hey, come on. Of course I could do that. The bird would love it. Twitchett, my terrier, old, stiff, blind, and deaf as she is, would be very excited. But I had a book to write.

I do take care of him—feed him, clean the cage, give him a bathing dish once a week (he only wants it once a week), and so on. I nursed him through an awful infestation of red mites. He shed all his feathers and I thought he was going to die, but modern chemistry came to the rescue in the form of a fumigant you hang in the cage forever—probably giving him liver cancer in his old age, but hey. The mites were hideous. (Luckily I had a great advisor in this crisis, Boston resident and friend Irene Pepperberg, of Alex the parrot fame.)

About the bird's old age—I assumed the canary would be a passing fancy, two years of responsibility at best. No, it seems canaries live 10 years or more—oof! But, it is easy for a distracted author to forget to check the bird's cage on a routine basis; I could neglect him, accidentally. Hmm.

As I sat down to tell you about my bird, I realized he has taken control of his own care. In addition to frequently bathing me with song, always sung deliberately in my direction, he also makes eye contact and chirps, loudly, peremptorily, from time to time when I'm passing by. As with my fish years ago (Chapter 13 in *Reaching the Animal Mind*), I have learned to respond. "You need something. What do you need? Ah, water is low. Seed needs refreshing. A little lettuce would be appreciated. Thanks for the cue, bird!" Click.

Lessons from Llamas

Karen discusses The Magical Manila Envelope and other tool fads that have swept the clicker training community.

Newcomers to operant training may place superstitious value on the specific tools they see others using, not realizing that it's the process, not the equipment, that counts.

Many zookeepers are now using clicker training to help their animals accept medical care, move back and forth from one cage to another, and generally fit into and enjoy their zoo lives more (See "Positively Healthy!" page 229 and "Zoos Then and Now," page 231). While different trainers prefer different tools (or sounds, like whistles or clicks), what's important to remember is that these cues are used only as connections to the animal's actions in real time. Cues provide information to the animal; with that information, the animal learns how to make "good stuff" happen. It's the timing and the use of information that's crucial, not one particular cue or another.

What do hippos want to hear?

I remember getting a frantic e-mail a while ago from a zookeeper who was responsible for three hippopotamuses. She needed to know—right now—where to get the kind of metal whistle that makes three sounds at once: a chord, a sort of hum, and a whistle. She wanted to train her hippos to come when called, go in their cages when told, and so on, and she knew hippos would only work for that unique sound! She knew that because she'd watched a hippo keeper at another zoo do wonders with his peculiar whistle (which he had probably bought at a garage sale, for all I know).

Hippos, of course, will work for any sound they can perceive. I think stamping on the floor or flashing a strobe light would work pretty well, too, whether the hippos were in or out of the water.

Strange cues

Remember that it's the process, not the equipment that counts. Still, everywhere in the agility competition world the target for teaching dogs to stop at the end of an obstacle and put a foot in the contact zone is the plastic lid from a margarine container. Someone, somewhere, needing a visual cue, grabbed one of those, and the rest is history.

In the dolphin world, the "jump" cue is a sideways sweep of your arm from low to high. It could be a clap or a bow, the dolphins don't care. But no, it's that sweep. In the zoo world, as operant training wends its way from keeper to keeper, there's a standard cue now for "open your mouth." What you do is put your thumb and fingers together, and then spread them wide: a mime of the lion's jaw, the hippo's maw, the gorilla's gape. And this cue has spread. All over the world, exotic animals and birds oblige by opening wide and giving the keeper a good look at the oral cavity. Any decaying teeth? Any signs of infection? Vets *love* to get that glimpse. Open wide, guys!

Llamas need cues, too

I discovered that I myself had triggered one of these contagious superstitions. When I lived in Seattle, Ellen Leach, a behaviorist and keeper at the Woodland Park Zoo, called me looking for help with the zoo's South American exhibit. There were several llamas in this exhibit, as well as tapirs and birds. The llamas had not been handled, and the keeper wondered if clicker training could make it easier to handle them, specifically to get them in and out of the barn when necessary, not just when they felt like it. I was glad to help, if I could.

I went to the zoo and visited the llama barn. The llamas were quite friendly, and promptly came in from outdoors to see me out of curiosity. But they were wary, too, since they had never been tamed or trained. Targeting would help, first to get them into the barn on cue and later to station them so one could proceed to touching, haltering, foot trimming, medicating, and so on.

One adult female was very receptive to clicks and treats, so I decided to demonstrate targeting with her. I looked around the barn for some harmless but unfamiliar object to use as a target (I didn't want to use a bucket or a lead rope or something with which she might already have a bad association). I saw an empty, used brown mailing envelope lying among some other papers on the keepers' desk. I picked that up and held it out to the llama. She looked at it: click, treat. She smelled it: click, treat. After a few more clicks, she was readily following the target—that old, brown mailing envelope—through the door to the outside and back inside again. Point proven.

Ellen Leach applied targeting, shaping, and the other principles of operant training to her llamas. Soon, I presume, the herd was manageable and the problems were over—at least they weren't asking me for more help.

My own claim to fame

A year or so later, there was a county fair near my house that included a llama festival, and I went for some fun. I fell into conversation with a man who'd brought some harness-trained llamas pulling two-wheeled carts, and he let me drive one around a pasture. It was just like driving one of my ponies, except there was no bit in its mouth. A light hand on the noseband was all that was required, and I have very light hands.

Later, I happened to see a substantial horse trailer arrive and start unloading more llamas. Guess what? Instead of leading them off with lead lines, the handlers were leading them off with...manila envelopes. Oh my goodness. What had I done? I hope that the manila envelope fad passed quickly, and that before they took their llamas out in public, trainers learned to replace targets with voice cues for loading or unloading.

Why Sheep Don't Herd Themselves

Like most of us, Karen has always been intrigued by herding dogs. Here's her answer to a letter from a retired doctor who posed a "clickerly" question about herding.

Dear Karen,

We just returned from a visit to New Zealand where we visited many sheep farms. We were shown several demonstrations where sheep dogs herd sheep in the direction indicated by a whistle signal from the dog handlers.

I asked one of the people demonstrating the dogs' skill why, if the sheep also hear the whistle signals, they do not automatically obey the signals before the dogs start to guide them. After all, they hear these very same signals over and over.

His answer to me was that sheep are not very intelligent animals and are deathly afraid of dogs.

I am not satisfied with his answer and ask you to help sort this out. As far as intelligence is concerned, we said the same about cats and pigs before we found ways to train them. So the intelligence factor may lie with humans; perhaps humans are not intelligent enough to know how to train the sheep directly instead of setting them up to respond to the barking and threatening moves of dogs.

We were also told that one good sheep dog could control up to 2,000 sheep. To me, this means that some of the 2,000 sheep may never see a dog but respond to her/his barking. So then, maybe sheep could be trained to respond as we wish by using sounds other than a dog barking, and dogs themselves are not actually necessary. What do you think?

Dr. William Reese
Sun City, Arizona

Dear Dr. Reese,

Thank you for your question about sheep dogs, sheep, and signals.

The question is not one of training but one of logic.

In fact, quite often neither the dogs nor the sheep know where they are being made to go. It might be to one gate or another, or it might be to a pen in the middle of the field, or maybe the shepherd just wants the dogs to gather the sheep and hold them for visual inspection.

So the whistles don't indicate destinations; they indicate actions. The whistles tell the dogs, principally, five things: go forward to the left, go forward to the right, come back (going left or right around the sheep), and stop where you are. There are other commands, such as slow and fast, but these five are the main ones. With these whistles, the shepherd moves the dogs around like chess pieces, and thus moves the sheep. Since you might be moving two dogs in different directions, often each dog may have its own individual whistles for each of the commands. (If you buy a new dog in New Zealand, you also get a tape of its whistles. If you lose the tape and forget a whistle, you'd better be able to reach the seller by telephone, or you're in trouble!)

In New Zealand, where the fields are huge, the dogs may not always be able to see all the sheep and vice versa; that is one reason New Zealanders use dogs that bark a lot. Bred to do so, they are called huntaways. In the smaller fields of, say, Wales, the dogs (called eye dogs) are silent, and can actually move the sheep just by glaring at them in a predatory way. In both cases, the sheep probably do learn that if they move away from the dogs, the pressure will ease. Therefore, they do not necessarily move with panic.

In the vast terrain in New Zealand, I have seen farmers team up and use as many as eight dogs at a time. Since the whistles are commands to the dog, not destination indicators, and since there are so many whistles that may change from time to time, the sheep have no opportunity to attach much meaning to any individual whistle.

In simpler situations, such as the smaller fields and flocks in England, you are right about sheep training themselves without needing or waiting for dogs to guide them. If there is one flock of sheep, and one frequently used gate to the field, the sheep can learn enough about dog whistles to steer themselves. Kay Laurence demonstrated that to me in England. She parked the car at a roadside pasture full of sheep. We walked over to the fence and Kay whistled a typical "command-to-a-sheepdog" whistle. Even though no dog was present, the hundred or so sheep in the field quit grazing and started calmly toward the exit gate, which was downhill near the farmhouse. They knew what to do.

Is that a more satisfying answer?

Cues as Information

Animals often operate from the perspective that "What you don't know could kill you." It makes their lives a lot easier if we can use a cue to let them know what's coming. Here's how.

We think of cues as something you must deliberately attach to a behavior, with a reinforcer to follow. "Sit" is the name for an act. If your dog responds to the cue by executing the behavior correctly, you will do something nice for him. That's a discriminative stimulus, linked to a behavior, followed by a reinforcer. Plain operant conditioning.

But you can also give cues that are purely information, not deliberately trained as antecedents to a particular response. For example, my dogs have a "wait" cue. It means, "Stop where you are, please." It's not a "stay." I don't make them wait interminably. It's not a behavior I deliberately trained, really. It's information about what I myself am going to do.

We go out the front door, and I realize that as usual I have forgotten the poop bag, or my scarf or gloves if it's cold. I say "wait" and duck back in the house. "Wait" means something on the order of, "She's going back inside and she'll be right out again, so we'll just stand here." At first I looped the leashes around the doorknob in a sort of token restraint, but after a few times, I didn't have to do that any more. Now I just drop the leashes.

The dogs don't take advantage of my brief absence to go sniffing around, or to look for squirrels; they just stand still. In fact they will stand still even if they are halfway down the front steps when they hear "wait." Twitchett, who is elderly and stiff, turns sideways on one step; Misha, who is long-legged and nimble, just freezes with his front end on one step and his hind end on another.

I use the "wait" cue when I am going to the car just to retrieve a package, not to take the dogs for a ride. And they stand still. I use the "wait" cue when I stop during a dog walk to chat with a neighbor, or when I'm fumbling for the car keys in a parking lot. It's handy, and a lot easier for the dogs than learning a formal down/stay, which in my opinion puts them in a very vulnerable position out in public (see "Why I Hate the Long Down," page 172).

Using the informational cue to overcome fear

A recent post on a zoo training list made me think about the wider training implications of this kind of cue, which, instead of telling your learner what to do, tells the animal something about what you yourself are going to do. A keeper asked the list for advice on crate-training a raccoon quickly. The animal was to be moved or shipped or something, and the keeper was given what looked like an impossibly short deadline.

Of course it's just like crate-training a dog; you shape attention to the crate; you shape approaching the crate. When the animal goes into the crate voluntarily, you start delivering the food inside the crate. As a good shaper, sometimes you ask the animal to stay in longer, sometimes shorter, before clicking, alternating between harder and easier, so things don't always get worse. One keeper had an excellent additional suggestion: from the very beginning, move the crate to different places in the enclosure so the animal focuses on the crate, and not on some specific location.

As we all know from experience, however, the real challenge comes when you start trying to close the door with the animal inside. Here's the suggestion that really caught my attention. Tell the raccoon what you are going to do. The raccoon is going into the crate willingly. Now you are ready to get it used to the closing of the door. Before you start, however, say "door." Then touch the door, click, remove your hand, and treat. Each time you start to reach for the door, give the cue that you are going to touch the door, stop touching the door after you click, and pay the animal for staying in the cage doing nothing.

The next step is to say "door," and move the door a few inches and back, click and treat. I'm sure you can see that you can quickly raise the criteria. Soon you will be able to say "door," shut the door, open it, shut it, open it, shut it, pick up the crate, put it down again, open the door, and then click and treat, with the animal staying calmly inside the whole time. After all, from the raccoon's standpoint, the animal is in total control! "All I have to do is sit here and let her do her door stuff, and she will click and treat."

This is *far* faster than desensitization, gradually escalating door moves and hoping the animal will get used to it. It is also faster than just clicking and treating for sitting still as the door moves, with who knows how much interior fear building up as the moves escalate. The cue tells the animal what you are going to do, and that vastly reduces the fear.

Also, as we know, cues are reinforcers, and this powerful aspect of the cue now works in your favor. Every click and treat makes your information about what you are going to do—mess around harmlessly with the door—also a cue to the animal to stay calm and stay still.

Thinking back to training mistakes of my own, I learned that teaching tolerance of confinement is an easy thing to mess up. If you rush the job, if "just this once" you are in a hurry, your half-trained, reluctant, and fearful animal (feral cat, shelter dog, unbroken colt, or whatever) may reach its breaking point and burst through the half-closed door. And *then* you have a huge re-training job ahead of you, because for days or weeks or more, just seeing the door move will remind the animal of the option of flight. Naming your own behavior protects you against that unfortunate circumstance.

So, where else could we use this cue about our own intentions? In her book, *Click to Calm,* Emma Parsons has a nice procedure for dealing with an aggressive dog. You see another dog coming, you fear your dog will explode, and you can't help tightening the leash. But suppose you have taught the dog that a tightened leash means that you are going to walk the other way and you will click your dog for coming with you. Rather than increasing the dog's anxiety, you have turned your own fear-generated behavior (stiffening, tightening the leash) into an informational cue: "We're going back the way we came now." Click, treat.

How might you use an informational cue about what *you* are doing in clipping a dog's nails? Getting a cat into its carrying case? Handling a green horse's ears? Try it out!

A Dog with a Catch

Some time ago, when Karen was giving a workshop in
England for about 20 of Kay Laurence's advanced students,
she was looking around for a dog to work with in a new
exercise about teaching cues. A huge, formidable
German shepherd was watching her.

"What about him?" I asked Kay.

"Yes, fine, all right," Kay said.

I trust Kay's dog sense, and this was her student so presumably she knew the dog. The owner, however, was dubious. "I don't think he'll work for anyone but me." Well, I don't want to step on your ego, I thought, but let's find out.

I ascertained his name and invited him over. He came at once. The dog, after all, had been lying there all morning watching other dogs earn clicks and treats. He'd probably been thinking, "I could do that," up to and including learning to jump rope as one clever collie had done.

What we were going to work on was transferring the cue for a particular behavior, from a word to a visual signal. What behavior did he have that was already on cue?

"Sit," I asked, and he dutifully sat. OK, we'll use that, I thought, and clicked and gave him a treat. The treats were tiny—a quarter-inch square of bologna each. Ouch. He was, as they say, hard-mouthed, and nipped my fingers accidentally, trying to take the treat. Kay can cure that in four clicks—I've seen her do it by shaping the behavior of taking the treat with lips and tongue rather than jaws and teeth; but I am not as deft and it takes me longer, which bores the audience. So, to spare my fingers when I find myself doing a demonstration with a dog that takes food roughly, I usually "feed the floor." After the next click, I dropped the treat onto the carpet. Wow! The dog caught it on the way down. He was a good catcher! So we arrived at another arrangement: "After each click I'll toss the treat, and you catch it." OK, said the dog.

I confess I had been a bit daunted by this very large and formidable-looking dog when he first stepped forward. Now, however, we were in business together, and I soon came to admire him very much. He quickly demonstrated that he could sit reliably on the verbal cue, so we proceeded to the second part of the lesson. I grabbed up a blue plastic bowl from the equipment table and showed it to him. This was going to be the new cue for "sit."

The way you transfer a cue is: Give the new cue, give the old cue, get the behavior, click, treat. Repeat: new cue, old cue, behavior, click, treat. I held out the blue bowl, said "sit," the dog sat, I clicked and tossed a treat. Work fast. Lots of clicks, lots of treats. After a few rounds of that, the next step is to give the new cue and pause. If you get even a whiff of the behavior, click, treat. At first, you're not aiming for perfection. Repeat that a few times, and the learner's confidence will grow. Soon the behavior will be offered on the new cue. In this case, after several repetitions of presenting the blue bowl and pausing before saying "sit," the dog hesitantly crouched on his hindquarters a little when he saw the bowl. It was the start of a sit, and I clicked it. A few clicks later the dog was sitting, promptly and confidently, each time the blue bowl appeared.

At this point the dog began sitting as soon as he'd caught his previous treat, whether he'd seen the bowl or not. "I know what she wants! I'll sit again!" The final stage would be to teach him to wait for the new cue: not to just sit on his own, in anticipation, before the cue was presented. I accomplished this by holding the bowl behind my back, and then taking advantage of a moment when he had just finished catching a treat and was still standing up and showing him the bowl. If he sat off-cue, that is, before I offered the bowl, I just moved a few feet away, so he had to stand up and move, too, to stay near me.

Then I presented the bowl again. If my timing was right, I presented the bowl during "standing up and waiting" behavior, thus rewarding that behavior. In effect, I was using the cue presentation to reward the dog for waiting for the cue to be presented, a quick and pleasant way to teach a dog to "obey" [wait for] a new signal.

Of course, I made mistakes. Sometimes I held out the bowl when I meant to put it behind my back, sometimes I clicked late, and so on. But the system is pretty forgiving. If most of the connections are right, you can get the job done. When I finally saw that the dog was deliberately controlling himself as he waited, crouching a little but keeping himself from sitting until he got the cue, I considered the mission accomplished. I ended the lesson with a big jackpot: I let him dive into a whole bowl of treats, for one satisfying mouthful.

All this time, as I tossed the treats here, there, and everywhere, the dog got better and better at catching. He could jump up as high as my head. He snatched his reward in midair with an audible chomp of those powerful jaws. He was having fun and so was I. It was a bit like training a killer whale. It's rather exciting to have all that energy (and all those teeth) in midair right in front of you, and yet feel safe.

A year later, I gave a lecture at the Association for Pet Dog Trainers (APDT) on the same exercise, how to transfer the cue. Before doing a demo with a live dog from the audience, I showed the video of this episode with the German shepherd and the blue bowl. From the camera's angle the shepherd's catches were awesome. Too awesome for one woman in the audience, who came up to me afterwards in the hall and said, scowling, "You taught that dog to jump up. You taught that dog to snap." No, he already knew that, I thought, and laughed, but I didn't say anything, and my giggle no doubt confirmed her judgment that I was an idiot.

I understood her distress, and I doubted any verbal explanation would help. What I had taught the dog, without thinking about it particularly at the time, was a controlled leap and a controlled chomp. Just exhibited at random, these are not nice behaviors, especially from a dog of that size and strength. In this session, however, the dog had learned to do them on cue. Trainer: Click. Trainer: cocks hand, ready to toss. The dog, seeing the hand motion, gathers himself. Toss, jump, chomp. It's a little behavior chain, under stimulus control (the hand position) and rewarded by the actual food-in-mouth. And what happens when behavior is under stimulus control, class? It tends to disappear in the absence of the stimulus. This dog was now far less likely to snap at food, or leap at people spontaneously, because he could control himself and do those things on purpose and on cue.

This traditional trainer was upset at seeing potentially hazardous behavior actually encouraged. In standard-practice training, one would try to "correct" such behaviors whenever they occurred, meaning one would aim to suppress them with punishment. I think my way of getting rid of random leaping and food snatching is more elegant. And we did it on the fly, so to speak, while the dog also learned something about controlling himself in the matter of waiting for cues.

Back in England, a couple of days after the blue-bowl episode, in a different lecture hall in a different town, that owner and that dog were in the front row. The dog clearly wanted to be up on stage, and this time the owner gave me permission with a big grin. When I called him the dog came galloping up the steps onto the stage, also with a big grin. He quickly demonstrated that he remembered his blue bowl, his sit, and his four-star, midair catch, too.

I might not have been quite so insouciant with him if I'd known his history, or rather, his lack of history. That night in a pub I found out from Kay that the shepherd was not a family pet; he'd been adopted two weeks earlier, and no one knew anything about him. Who knows what event had made him homeless or what misbehavior might lie in his background? He had come to trust his new owner a little, but she was quite right when she said she didn't know if he'd work for someone else. Clicker training, however, doesn't require personal trust; it builds it. In two directions: dog to person, and person to dog. And that certainly had worked, for both of us.

Waving Goodbye to Sloppy Cueing

In the business of cues, half the battle is training
the animal to respond; the other half is training yourself
to give the cue consistently.

Often dogs fail to respond to a cue not because they are being stubborn, or because they don't know the cue, but because we gave the cue carelessly. With the wrong hand, with another word or two mixed into it, or in a new environment where some aspect of the cue that the dog relied on is changed or missing. If you've been inconsistent, and the dog doesn't always respond even though the behavior itself is well trained, transferring the cue you've been using to a new cue can help.

At the annual meeting of the Association for Pet Dog Trainers (APDT), I was giving a workshop on cues and cueing. To show how to transfer a cue, I asked for a clicker dog with a big repertoire of on-cue behaviors. An audience member volunteered her dog. I asked "What's the dog's favorite behavior?" "The wave." Perfect. Up on the stage came the APDT member and her nice herding dog. I clicked and treated the dog; he focused on me instantly. (For a clicker dog, finding a new person who clicks must feel the way it feels for us when we're in a totally foreign city and run across someone—anyone!—who speaks English.)

"Will he do his wave for me?" I started to ask, and the audience laughed. The word "wave" was the cue, and the dog had his paw over his head before I finished the sentence. Good! Communication had been established. So in roughly five minutes I transferred the cue for wave from a spoken word to my sunglasses. Show him the glasses, he waves.

Well, why would you want him to do that? Maybe for a joke: "I'm going to the beach, who wants to come?" I said, taking out my sunglasses and putting them on. "Me, me," says the dog, waving his paw in the air.

More importantly we had demonstrated that anything can be a cue, including an object; the dog had no preferences. We had also seen that you can transfer to a new cue—a hand signal or an object instead of a voice signal, say—in next to no time, if you do it right (See "A Dog with a Catch," page 195).

Next, I transferred the cue to the owner. At first, when she picked up the glasses and showed them to the dog, the dog practically shrugged. The cue was meaningless. Why? I studied the situation and realized I'm left-handed and held the sunglasses in my left hand. The owner had picked them up in her right hand. Good example of differences in a cue that might be small to us, but huge to the dog! Then she moved the glasses into her left hand and showed them to the dog, and Bingo—big wave.

In three or four more clicks the job was done. Could she have gone on then to transfer the cue to her right hand? Of course. Right hand—pause—left hand, behavior, click, treat. Fade out the left hand. Might take no more than four clicks, that way; retraining it from scratch in the right hand might take thirty. Save yourself trouble. Transfer cues step by step.

Benefits of a new cue

I've observed that transferring a well-learned behavior to a new cue cuts down on the time needed to get the dog performing well in new, scary, and distracting environments. In fact, some people are using this technique to reduce or eliminate the need for elaborate desensitization and gradual exposure to new environments such as show grounds and city streets.

Fading the Click?

For everything there is a season,
A time for prompts. A time to fade them.
A time for cues. A time to fade them.
A time to click. A time to be still.

The first step is realizing when you're using prompts and
cues. Karen gives a primer here in the art of introducing and
fading cues and prompts.

"When do I fade the click?" "How do I fade the click?" We hear those
questions all the time. The smart-aleck answer is "Never." Because we don't
"fade" the click. Fading means doing something smaller and smaller until
a tiny version of the original stimulus will serve, or until the learner no
longer needs outside help to do the behavior. We don't do that with a
click; either you clicked, or you didn't. Period. The term "fading" applies to
prompts and cues, not to the marker signal.

Fading a prompt

Here's an example of fading a prompt. One of my favorite shaping
demos consists of teaching a dog to walk backwards. I usually start by
leaning toward the standing dog, which usually makes him lean away
from me or take a step back. Click. I've prompted the behavior; he didn't
do it on his own. I may do that again, two or three more times, clicking
any hind foot movement. But then I will "fade" the prompt. I'll lean just
a tiny bit and hope that he will voluntarily step backward. If he does,

click! Treat! Then I won't lean at all and hope that he'll repeat the behavior without the prompt. Then we can progress to two steps backward, then three, and so on.

When you are shaping behavior, the good trainer is extremely still and quiet. From the moment I've faded the prompt I'll be very careful not to give him any body cues at all—or facial cues or eye contact. I don't *want* the dog attending to my moves now, or he will become "prompt-dependent," always looking for messages from me to tell him when to move. When you're working on a new behavior, don't cheerlead or encourage your dog with gestures and words; these are meaningless signals you're just going to have to fade out later. Let the dog figure it out in peace.

At ClickerExpo in 2006, I did the back-up shaping demo on stage with a big Samoyed named Max. I chose Max because I thought the audience could easily see him take a step backwards with his big furry white hind paws. While the owner stood behind me holding the loose end of Max's leash, I faced Max and leaned over him a little. The prompt worked; he moved back a step. I clicked during the move, gave him a treat, called him forward, and did it again.

I could easily watch one back foot moving but when I raised criteria to two, three, and four steps, I ran into a problem: I couldn't see the other foot because Max's big furry white body was in the way. Luckily my co-worker Bill Peña was running a follow-camera on me and the dog, so people in the back of the room could see the demo on a big screen next to the stage. Looking over my shoulder at the screen I got the audience's view of Max, too. So I did the rest of the training facing away from Max, watching and clicking his foot movements on the screen, and handing him each treat from behind my back. That *really* kept me from giving Max any more prompts. He learned quickly to back up as far as space permitted, getting his clicks and taking his treats politely. I never even looked at him. (If you are having a hard time not coaxing and encouraging your dog with words and wiggles, try shaping in a mirror!)

Fading a cue

So, we fade prompts. We also fade cues. Suppose I select a dog for the back-up demo that is very swift and learns quickly to back halfway across the room. I might now want to add a cue. I could start by leaning forward again while making a pushing gesture with my hand. I would click the first step backward, not waiting for the whole behavior I've just shaped. I'm now using the click to shape the response to the cue, not the behavior itself.

On the next trial, I might omit the leaning and give him just the pushing gesture; and if that gets a response, I might click the third step backwards. When he's backing up continuously until he hears the click, I would teach him to wait for the cue, and not just start backing up without it. When he'll do that, I can assume he understands the new cue.

I could now fade the cue even more, by making the gesture smaller and smaller on each try until the dog is rushing backward at a tiny move of one finger. Fading cues is an important part of developing a performance dog for stage or screen or in freestyle dance competitions. The audience sees the amazing behaviors, but never sees the small moves that tell the dog what to do.

Making all your cues as discreet as possible forces the animal to focus sharply on what you say and do, which heightens the accuracy of cue response. I think the fact that the animal is focusing hard and waiting for your information also intensifies the reinforcing effect of the cue. And fading your cues has another benefit: it disciplines you to resist turning learned cues into prompts, yelling louder or waving harder if the animal doesn't respond. (If the animal doesn't respond to a cue, of course you need to work some more on criteria involving that cue; making the cue bigger won't fix the problem.)

Dropping the click

The click is a whole other matter. To eliminate the beginner's anxiety, of course you don't have to click forever. The click is for teaching behaviors and cues. During the shaping process we gradually click and treat less often as we require more and longer episodes of behavior to earn a click. Then, once the learner knows what to do and when to do it, for many behaviors you don't need to click any more; a nod or a smile or a word can tell a dog he's doing fine.

Life provides all kinds of reinforcers for dogs that know what to do. Doors open, people scratch you in just the right place, you get walks, rides in the car, dinner, and so on. In these daylong exchanges, there are many opportunities to make nice results contingent on nice behavior. Maybe you once had to teach your dog his everyday skills in small increments with clicks and treats, but now they are part of life and can be maintained with looser and more general pleasant events.

Furthermore, continuous unnecessary clicks can actually be harmful. The click is a powerful way to convey *new* information to the dog. It should make the dog feel as if he's solved a puzzle. "Ah-ha! That's what they want!" People who experience TAGteach, being taught some skill or move with the marker signal, often report a sensation of winning, a feeling of excitement when they hear the tag, or click. That's an important part of the reinforcing role of the marker.

In fact, once the behavior and the cue have become routine, you probably *should* stop clicking. During lecture sessions at ClickerExpo, I sometimes see attendees clicking and treating dogs repeatedly just for lying down and being quiet, when the dog is already lying down and being quiet anyway. Using the click in this way, just to maintain behavior that's already been learned, may actually devalue the click. TAGteacher Theresa McKeon tells me that too much clicking, especially for already-mastered skills, is actually annoying to her gymnastics students. For TAGteachers, the optimal situation appears to be short intensive periods of tagging for new skills, alternating with longer sessions of practice, games, or other exercises using already-learned skills and reinforcing very sporadically, and with social praise, free time, and other more general reinforcers.

Attila Szkukalek and Fly: High-level performance "post-click"

Here's another question that outsiders often bring up: "I can see how you don't always need to click around the house, but what about something hard? How about obedience trials? How can you clicker train for that? You can't click and treat in the ring!"

That's true, but you don't need to, because a) your dog's behaviors are already trained and b) you can maintain the behaviors with other reinforcers: your well-taught cues. The whole exercise is, or should be, a chain (see "Making the Connection: Behavior Chains," page 208), with each behavior continuing until the handler gives a cue for the next behavior. Properly trained, the dog must be thinking all the time, looking for the next cue; and each cue reinforces the behavior that was going on when the next cue was given.

The results can be spectacular. Attila Szkukalek is a superb freestyle artist, an actor as well as a dancer; and so is his dog Fly. Attila says on his website (www.dogdance.net), "The majority of freestyle dogs are clicker trained. Since the artistic impression is judged, handlers cannot use compulsive training methods that would subdue the dog's enjoyment and spirit."

But you'll never hear a click during Attila and Fly's performance; that part is over and done with. Other reinforcers now maintain the extraordinarily complex sets of behaviors. Take for example Attila's charming "Charlie Chaplin" routine. By my rough count, the dog performs at least 40 different cued behaviors in about 3 minutes. Some of these are long-duration, such as scooting backwards 20 yards or more; some are carried out multiple times, such as dancing in a circle on the hind legs around the human dancer. Some reflect incredibly refined shaping skills—for example, in some moves the dog's facial expression is actually a shaped part of the behavior.

The dog is eager and excited, watching for each cue and responding instantly. What are the cues? They include voice, hand signals, targets (the cane and handkerchief, which are also props in the "story"), and leg, arm, and body moves galore. Each cue is a reinforcer, of course, but there are other reinforcers, too. In his writings about freestyle, Attila mentions specific "favorite" behaviors that are highly rewarding to the dog and can be used to reinforce behaviors a dog finds difficult, such as distance work. In training, Attila also uses the occasional high-value treat. By the way, I have seen Attila perform in England many times now and I've watched his videos assiduously. I defy you to spot any moments in performance when Fly gets a treat.

To make things even more challenging, the performance is also sprinkled with moves and moments that are *not* cues for the dog, such as the actor/dancer's facial expressions, gestures for the audience's benefit, and the music. It takes a lot of discipline on the trainer's part to keep all this straight!

Attila is a biochemist in his day job. He hails from Czechoslovakia originally but now, with his wife and children, lives and works in England. Teaching, training, and competing in freestyle are his "addictions," as Attila puts it.

Making the Connection: Behavior Chains

If it's hard to wrap your head around the concept that a cue can

also act like a click, welcome to the world of behavior chains.

Karen will guide you through it.

A behavior chain is an event in which units of behavior occur in sequences and are linked together by learned cues. Back-chaining, which means teaching those units in reverse order and reinforcing each unit with the cue for the next, is a training technique. We use this technique to take advantage of the intrinsic nature of the event.

The cue as a reinforcer

The key to understanding what's going on in a behavior chain—and why it creates reliable behavior—is to know that a cue is also a conditioned reinforcer, like the clicker. Put another way, a cue, which is the "green light" for a clickable behavior that leads to some kind of treat, becomes in itself a good and rewarding event. By carefully timing the instant in which you give the cue, you can reinforce some other behavior that's going on at that time. In training a behavior chain, you can mark a behavior and reinforce it and cue the next behavior simultaneously.

Example of a behavior chain

There are many behavior chains in everyday life. When I take my dog out of his crate in the morning, I immediately take him for a walk. This involves many little units of previously learned behavior: standing still to have his leash put on (instead of romping and playing, as he'd like to do), waiting politely at the door, and again, if I ask for it, at the top of the porch stairs—I often have to go back for gloves or some other forgotten item—then walking without pulling, and so on. Each of these behavioral units was taught individually at first. Now they are linked, with each cue reinforcing the previous behavior. For example, when he's waiting quietly at the open door I reinforce that by saying, "let's go." The actual reinforcer is the walk itself. The cue to go through the door reinforces the polite waiting.

Here's the important point: when I was developing waiting politely at the open door, I didn't need to click and treat for the wait before I said, "let's go." I didn't need to say, "good boy" as praise for the wait, before I said, "let's go." Unless the known cue, "let's go" in this case, is also associated with punishment, it is in itself a powerful positive reinforcer (for more on this topic see "Poisoning the Cue," page 216). It can function to mark the behavior, just like a click.

In building my go-for-a-walk chain, and all the units inside it, I got the job done just using cues as reinforcers in the natural course of our daily routine. Life offers many opportunities—going into and out of the car, visiting other houses, and going to the vet—for indicating and reinforcing behaviors that add up to "good manners" by using the cues "wait" and "let's go" in a timely fashion.

What if I had a dog not yet attentive to learning new cues? If I were, for example, temporarily taking care of someone else's rambunctious, clueless door-dasher, a few contingent cues, even if they are conditioned reinforcers, might not be enough to get the job done efficiently. In that case, I would certainly bring out a clicker and treats to put in a bunch of brief but intensive C/T sessions dedicated to learning that "sit" at the door is the only way to get the door to open; to learning that waiting for the cue "let's go" is the best way to get permission to move forward, and so on. I would also want to make sure that I myself did not carelessly "break" the chain, for example by putting the leash on, sitting the dog at the door, and then going off to make a phone call or something, leaving the dog unreinforced for the front end of the chain.

Back-chaining

Many of the behaviors we train our dogs to do are really behavior chains. Heeling, retrieving, running an agility course, almost all obedience exercises, tracking, gaiting, and stacking in the show ring—all are chains. While the various units of a chain can be trained individually in no particular order, linking them together is far more easily done if you work from the end of the chain, in reverse order, toward the beginning.

Back-chaining the retrieve with the clicker

Start with the drop or give. Establish a cue for that, then back up to the take, hold and give, and then the take, carry, hold, and give. Train "go over and find it" with the object stationary on the floor, after that. Last of all, introduce the throw, watch, and chase (or chase and catch, for Frisbee) part of the retrieve. Doing these steps with clicks and treats is fast and fun. The steps can be taught to puppies as soon as they can see, hear, and totter about on four legs. If you back-chain the retrieve you will always have a zesty, eager partner who will never try to play "keep-away" instead of fetching the object back to you.

With back-chaining, you start with the last item in the chain—in the retrieve, it would be the give. You shape that behavior, put it on cue, and then insert the next part: hold until I say "give." Then you back up one more step, and teach the take, first from your hand, then from the ground. Building the chain backward ensures that you are always moving toward reinforcement—the prize at the end of the chain—and that each part in the chain is strengthened, every time, by the cue for the next part.

In building a behavior chain or inserting new behaviors into the front end of a chain, you don't need to click and treat every unit. Direct reinforcement of the new behavior may not be necessary, since you are already using the next cue as a click. Continuing to the next link in the chain is more reinforcing than interrupting the chain with a minor reinforcer, such as a food treat. Going toward a known way to success can be so important that the dog would rather keep working toward the goal than stop to eat or to acknowledge praise.

Uses of back-chaining

If you are starting to build complex chains for competition, you will go faster and your dog will understand better if you build each unit separately, join the units up from back to front, and practice the chains over and over—always in groups of units rather than running through the whole chain every time. For example, in the articles exercise in utility, one might build and practice the mini-chain of "pick up, bring, hold, give," separately from the mini-chain of "go out, find, and select the right article." Of course you could also occasionally practice the mini-chain of "select, pick up, and bring," clicking as the dog turns back with the right article, then going to him with the treat, to reinforce his good selection quickly. A side benefit of training in mini-chains is that if one unit goes wrong in performance, you can take that chunk out, shape that unit and its associated behaviors again, and then put the mini-chain back into the long chain. Of course, each of these mini-chains should also be built backward, as in "Back-chaining the retrieve with the clicker" above.

Skills that benefit from back-chaining include the retrieve, tracking, search and scent work (start with the reporting behavior), and any performance task that happens at a distance, including field trials and herding. Incidentally, it's not just dog performance that benefits from back-chaining. If you ever have to memorize a piece of music, or a poem, or a speech, or a dance routine, it will go much faster if you break it into little chunks and learn the last chunk first, then the next to last, and so on, backing up to the start.

Prop cues

A common misconception is that a behavior chain is a series of behaviors that are initiated by a single cue. In fact, that's the way some behaviors "look" to us, because we tend to ignore any information the dog gets that does not come directly from the handler. Take, for example, the obedience exercise of retrieving the dumbbell over a jump. Some dogs whip through it with accuracy and panache. It certainly looks as if the dog has memorized the whole sequence and is doing it on a single cue, the owner's send-out from the starting position of sitting at heel. However, this cluster of behaviors is riddled with object-related cues, or what the bird trainers call prop cues.

The initial unit, leaving heel position and taking the jump, is cued by the handler. The sight of the dumbbell on the other side, however, is the cue for picking up the dumbbell, and also the reinforcer for taking the jump. The feel of the dumbbell in the mouth is the cue to turn back to the owner (taking the dumbbell home) and then, when the dog turns back, the sight of the jump is the cue to take the jump—and the sight of the owner standing there in a particular pose reinforces the jump and also cues the "front" behavior, and so on.

What if the dumbbell isn't there when the dog gets over the jump? What if it took a bad bounce and went out of the ring? That can happen. It's the rare dog that turns and takes the jump back anyway; mostly the dogs just wander around looking confused. There is no cue (sight of dumbbell), so no pickup behavior occurs. No cue (dumbbell in mouth), so no turn-and-jump behavior. The loss of a cue in mid-chain is not the only way a behavior chain can go to pieces, but it's a common one.

What can you do to train against the mishap of a dumbbell bouncing out of sight? Here's one recipe. Teach the dog to hunt for and find the dumbbell by scent, for a click and treat, indoors, around the house—and then outside, under furniture, in clumps of grass, under ring gates. Establish that if the dumbbell is in sight, pick it up and bring it; if it's not, find it by scent, then pick it up.

Patterning

Some people maintain that the best way to get "reliability" in performing a series of behaviors is to train with many, many repetitions of the same sequence over and over, sometimes called "patterning." It's hard, it's boring, and the resulting behavior is very vulnerable to changes in the environment. Sometimes, however, it seems to work. Why? In fact, if the sequence is holding up, it's probably not because of the many repetitions, but because there are cues within the chain that are reinforcing the pieces of the pattern. We just don't recognize them as cues because they are environmental; they aren't deliberate words or signals from us.

Different kinds of chains

Repeating a single behavior

Even some very experienced trainers consider that a behavior chain can only consist of a series of the same behaviors repeated over and over. That is one kind of chain. For example, running a horse or dog down a jump chute over a series of identical jumps is a chain; the sight of each jump is the reinforcer for the last jump and the cue for the next one. When the jumps stop, the jumping stops too.

Many behaviors, always in the same sequence

Some canine sports, such as flyball, involve a variety of behaviors that always occur in the same order. Freestyle, heeling to music, or dancing with dogs (theses terms are synonymous) is another example. Routines are choreographed and performed in a given sequence. That's a behavior chain. Each cue, whether an object cue—a jump in front of you or a handler's cue, a word or movement—signals the shift to a new behavior and also reinforces the behavior that is going on simultaneously.

In cases where the animal appears to know the sequence by heart, very often he is still responding to cues, too. They might be position cues: we always canter when we reach this end of the arena. They may be superstitious cues from the handler such as weight shifts of which the handler is unaware. Or they may be environmental cues, such as music or jumps. The result is still a behavior chain.

There are intrinsic hazards in building a chain that will always be performed in the same sequence. If the animal actually memorizes the sequence—"First I always do this, then that, then the other"—he may begin doing it on his own, anticipating the next behavior. When the animal "jumps the gun" (called anticipation) and acts without the cue (a common occurrence in roping horses), behaviors inside the chain fail to be reinforced, and start to break down. We see this happen frequently when training for Flyball competition and with the Drop on Recall obedience exercise. It's vital to deal with anticipation immediately, retraining that unit or cluster of behaviors to make sure that the animal waits for the cues; otherwise, problems will multiply.

Flexible chains

Of course the biggest and most important chains in dog training are the performance chains: long sequences of many behaviors, linked, reinforced, and thus maintained by cues, in which the individual units may come in virtually random sequences. Running an agility course is an example. The performance of obstacles occurs in a continuous stream, but the obstacles may be in any sequence and in any location. Running the course is a flexible chain, and one in which the function of cue as reinforcer is particularly obvious.

Take, for example, the challenge of contact zones. Some obstacles, such as the A-frame and the dog walk, have contact areas at the start and finish. The dog must touch those contact areas on the way up and again on the way down. The requirement keeps the dog safe; if he passes through the contact areas correctly, he can't jump onto the obstacle from a bad angle or bail out early from too high up, risking injury.

Because the course is different in every trial, every time the dog takes one obstacle, the handler has to give a cue to identify the next obstacle.

Common sense might lead the handler to wait until the dog completes one obstacle before telling him where to go next; but common sense is wrong in this case because of those contact zones. If you habitually give the next cue when the dog is already on the grass, guess what. He's going to start leaping over those contact zones to get on the grass because that is where he is reinforced with the cue for the next behavior. If you always give the next cue while the dog is in the contact zones, you reinforce being in the contact zone, and the dog will be certain to hit that spot.

It really doesn't matter what sequence the obstacles come in, but it does matter very much when the handler gives the cue. If the cue comes late, you have lost the opportunity to reinforce the previous task with precision. And if the cue comes way too late, so that the animal meanwhile acts independently and goes off on its own, you have broken the whole chain. All the previous behaviors are now at risk, especially if this event is repeated often; it is not the dog (usually assumed to be easily distracted), but the handler's timing of the cues that is at fault.

The linking of behaviors by well-timed cues is the essential factor in maintaining "reliability" in all long, complex, flexible chains. This includes obedience, tracking, search and rescue, field trials, hunting, retrieving, service work, and police work. I don't mean that the dog shouldn't work on its own initiative; of course it must, but always under direction as well. When the work is "on cue" the chains stay reinforced—because the cues are reinforcers.

Poisoning the Cue

Punishment is pernicious; it has a way of creeping in and not only weakening and disrupting behaviors, but also totally changing the dog's attitude toward the trainer and the training, as this story illustrates.

The Association for Behavior Analysis (ABA) is the organization for the "parent" science of clicker training. At its annual meeting in 2004, Nicole Murrey, a University of North Texas graduate student, reported on using her own family dog, an eight-year-old poodle mix, to develop and demonstrate a poisoned cue. She laid out a grid on the kitchen floor, so the dog's whereabouts could be clearly identified. She videotaped every training and testing session. First, using clicker and treats, she taught the dog to go out to a particular spot and then come back when called, on the cue "ven," Spanish for "come." We saw the usual clicker response: a fast, straight, merry, tail-wagging "come." Then Nicole also taught her dog to come, with a click and treat on arrival, but added a correction—a leash tug on the harness—if the dog was slow, or didn't come all the way. The cue for this was "punir," the French word for "punish." The dog did learn to come on that cue, but slowly, dawdling or wandering around before starting the behavior, and with a totally different carriage and demeanor.

As we know, positively trained cues can be powerful reinforcers. Nicole then tested each cue's reinforcing powers by using it in place of a click to mark a new behavior. She first taught the behavior of going to a mark on the floor on the left side of the room, for a click and treat. She then reversed that, teaching the dog to go to a mark on the right, using the ven cue as the reinforcer instead of the click. The dog learned that new task in about three trials; we saw it on the video. The dog was working very fast, tail wagging a mile a minute, mighty pleased with itself. "Go to this mark, hear the "ven" cue, rush to my owner, get my click and my treat." Nicole then did the same training—go to a new spot—but with the "punir" cue as the reinforcer. Wow! The behavior did develop, but it took forever, the dog never fully learned the behavior, and we saw lots of "distraction," "forgetting," and "not paying attention" going on—the things dogs show us in many a training class. (Do we treat the symptoms and not the cause? I believe we often do.)

This presentation raised a lot of questions from the audience, you may be sure, and Nicole handled them calmly and graciously, standing her ground, for example, on why she used a harness instead of a collar. "The dog is small, the dog is old, and I didn't want to risk hurting her by pulling on her neck." Well my goodness, one audience member argued, lots of people have small dogs and still use buckle collars—and leash corrections! "I didn't feel comfortable about doing that," Nicole said firmly. Click.

The Limited Hold

"If you click slow sits, you get slow sits."
Karen discusses the niceties of reducing response time
in animals—and in people.

The limited hold is scientific terminology—laboratory slang, really—for a good way to use the marker and reinforcer to speed up the response to a cue. We're all used to sluggish responses. You call folks for supper, and in due course, they come; meanwhile the soufflé falls or the soup gets cold. You ask the class members to be quiet, and some sit down and shut up, but it's quite a while before the last few stop talking. You call your dog to come in the house and it comes, grudgingly, finding half a dozen new things to sniff before actually reaching the back door.

Suppose you are dealing with just such a behavior. You give the cue, and you get the response, but after a delay. Now you want to fix that. First, practice it a few times to judge the average length of the delay. You can count seconds to yourself, or actually use a stopwatch. Once you have a measurement of the average time, make that your time limit. Now give the cue, watch the time go by, and reinforce the response if it occurs within the average time. If the cue response occurs over the time limit, call an end to the trial (with dogs, easily done by moving to a new location). Then give the cue again and start your countdown again.

In *Don't Shoot the Dog*, I told this story about a limited hold. At Sea Life Park in the 1960s, one of our most effective show highlights was a group of six little spinner dolphins that did various leaps and whirls in response to underwater sound cues. The most spectacular behavior was the aerial spin for which they are named. Initially, when that cue went on, spins occurred raggedly and sporadically across a 15-second period. Using a stopwatch, we started turning the cue on only for 12 seconds, and marking and rewarding only spins that occurred during that time. When most of the animals were spinning within that period we cranked down the limited hold, shortening the time to 10 seconds, then 5, and eventually to 2.5 seconds. It couldn't go much shorter, because the animals had to dive first to get up speed to jump fast enough to do a good spin. In any case, every animal learned that in order to get a fish it had to hit the air and perform the spin within 2.5 seconds of the time the cue went on.

As a result, the animals poised themselves attentively near the underwater loudspeaker. When the spin cue went on, the pool erupted in an explosion of whirling bodies in the air; it was quite spectacular. One day while sitting among the audience I was amused to overhear a professorial type firmly informing his companions that the only way we could be getting that kind of response was by electric shock.

Accelerating sits

In the lab, the length of time your learner takes to respond to a cue is called "latency." A long gap is called high latency; a quick response shows low latency. You can sometimes shape a low latency response, without bothering to measure the time, by asking for a lot of responses in quick succession and reinforcing only the quicker ones. For example, in demonstrating a shift in latency, I sometimes seek out a dog with a slow, lethargic sit: the sit occurs after the cue, but with high latency. I call him, back up a step or two (to get him moving forward), click and treat when he comes, and repeat a couple of times. When the dog is coming with me willingly, I back up and say "sit," stop, and click just as his back legs begin to fold, and treat. Call him, back up quickly, and cue "sit," again clicking the *act* of sitting, not waiting for the full sit. Then I progress to backing up, cueing the sit, and clicking as the rump hits the ground; and then to clicking as he sits but only if he sits instantly when he hears the word. If he hesitates, I back up again, call him, and cue "sit." In about 20 clicks (and 30 seconds or so), the limited hold is down to practically zero and the dog is sitting like a champion. One would then repeat this facing the other direction, and then perhaps in another room, on another day, and finally outdoors, to generalize the low-latency response. And, of course, having embarked on this repair job, you would also drop from your own repertoire the habit of reinforcing slow sits.

Speeding up recalls

With the "come" cue, often a dawdly behavior, it can be hard to gauge what's faster and what's slower just by watching. In this case, a methodical application of a limited hold can be useful. Think of the limited hold as a single criterion, like height of a jump, duration of a sit, strength of a push. You can train it in one situation, and then extend the criterion to other times and places. So, in the example of a dog that sniffs and dawdles over every blade of grass en route to the back door, you might train a low-latency recall indoors, first, and then gradually add speed as a criterion of the recall, in other circumstances. A long hallway is a great place to do this. Mark a chalk line across the floor at each end of the hall. Stand behind the line at one end. Enlist a helper to call or lead the dog back to the other end of the hall between each run, or to hold the dog while you move to the other end. Make the run very short (five feet or so) the first few times, clicking the dog's arrival at the chalk line and giving a highly preferred treat.

Now extend the run to the length of the hall. How are you going to tell which of two similar runs was faster? Most of us recite the alphabet at a pretty steady rate. You can use that to measure small increments of improvement. "Come," you say, and as the dog shambles toward you, you recite a-b-c-d-e-f-g-h-i-j-k-l until he crosses your chalk line, whereupon you click and treat. On the next run you'll know if he speeds up a little, because he will reach you by h or j instead of k. Good! Click/treat. And, of course, if he takes longer, you'll know that too.

Side effects of low latency

The object is not to punish slower runs, but to pay for faster ones. You make your criterion roomy enough so that most, but not all, of the runs are within your chosen limit. The procedure itself may naturally speed up the dog. As he begins to pick up speed you can introduce the limited hold: choose the maximum letter you'll tolerate, and if he doesn't get there by that letter, change ends and call him back the other way. I have seen the world's slowest Newfie, who plodded all the way to l-m-n-o-p on his first try, end up by responding to the "come" cue at a nice canter, getting to me by b-c-d, after about 15 clicks and treats.

Oh, of course this works with people. When I'm lecturing with a new group of people, for example, at the very beginning I ask for quiet with a gesture, a raised hand. Generally everyone goes on talking. Then I get down from the podium and go around the room for about 30 seconds, clicking and handing a treat (Hershey's Kisses) to any people who are being quiet. The next time I ask for quiet, usually when starting up again after a break, I stand still and look at my watch. When most of the people have stopped talking, I click into the microphone and say "Good!" And start talking. By the third time it happens, the audience falls silent, except for a giggle here and there ("She's training us!") as soon as I step up to the podium and raise my hand. I've established a cue, the raised hand, and I've also shaped a good low latency response.

As Bob and Marian Bailey put it, latencies are contagious. If you are attentive to selectively reinforcing brisk responses to cues in a few crucial responses—paying only for low latency responses—all of that learner's cue responses will tend to be brisker. That makes for a sharp-looking worker! On the other hand, if you generally accept and pay for any eventual response to your cue or request, even if it took the learner (the dog, the child, the teenager, the spouse) forever to actually do it, then high latencies and long waits are what you'll generally get. One example I've experienced personally is the difference between getting on one of the guest riding horses at a dude ranch, and getting on a working cowpony, a cutting or roping horse. The horse in the riding stable starts up into a slow walk after you've kicked it a few times, and requires several kicks and some urging to break into a trot. Steering it may also require some effort, and it stops slowly too, going from a canter down to a jog and finally a slow walk again. In contrast, the cutting horse turns on a dime; it moves and changes speed and direction instantly, and on the smallest of cues. What a pleasure; the latencies are so low it feels as if all you have to do is think what you want the horse to do, and it's already happening.

And all you need to do to have your own learners respond that way is to value and reinforce quick responses to cues, withhold reinforcement for slow responses, and, when the difference is hard to measure, reach for that useful tool, the limited hold.

Did He Say "Left" or "Right"? Descriptive Cues

A search and rescue dog that understands directional
signals can be remotely controlled—a huge advantage in
rough terrain. A clicker-savvy dog can learn these kinds of
cues quite quickly. An experienced clicker trainer got the
rudiments down in eight, five-minute training
sessions with a puppy.

At ClickerExpo, people are always impressed to see how often speakers sit in on each other's sessions. Well, we learn new things from each other. So it's quite a thrill. One of the many examples was Ken Ramirez' session in 2004 on how to train cues that are really adjectives: left/right, big/small, near/far, numbers, colors, and "sentences"—cues that can be used individually or strung together in series. Every teacher who wasn't actually speaking elsewhere at the time was in that room, and so were members of the staff and most of the advanced trainer attendees—from the Tennessee FEMA Team to Australia's four-time national Field Trial Champion.

Researcher Irene Pepperberg's parrot, Alex, has become famous, Ken pointed out, for being able to identify descriptive details, such as the shape, color, and materials, of various objects. Ken showed us some Nova footage of that historic set of experiments. Meanwhile, Ken and his staff at the Shedd Aquarium are using similar descriptive or modifier cues, not just directional cues, with marine mammals. Ken also has taught the uses of modifier cues to many search-and-rescue dog handlers, allowing them to steer dogs at a distance in places where a person can't go.

Then, step by step, Ken took us through the training process. He told us what you need in prior training: an animal that's clicker-wise, has a solid retrieve, understands many cues, and can play the "creative" game. He gave us some advice about syntax—in other words, in what order to give multiple cues (back-chain them).

In fact, in the three days just before coming to ClickerExpo, using a friend's 10-month-old spaniel pup and hand signals for cues, Ken trained and videotaped (and showed us) the following behavior: "Look at this toy (a stuffed mouse). Now, taking the left path around the furniture, go get the matching toy (from a pile of toys in the next room), and bring it back." Dog spins around, zooms off to the left around the ottoman in its way, disappears, and comes back at about a million miles an hour with an identical toy in its mouth. Next, Ken held up a fancy toy ball and cued the dog to take the right-hand path and bring the toy that matched that ball. Done deal.

It took eight, five-minute training sessions. We saw key moments of each one. Ken used only petting and praise as reinforcement, no clicks, no treats. Mind you, the dog really *liked* his petting, and Ken's impeccable timing allowed him to use face and voice as marker signals with perfect clarity. Most of us mortals might be better off to stick with the click. However, we all got the picture. And we all had the same reaction. "Gotta try that!"

As I write this, I've only been back from ClickerExpo for three days, and I haven't heard from *everyone* yet. But I've heard from some. Sherri Lippman and Virginia Broitman are already stacking modifier cues with their dogs. Alexandra Kurland is training matching-to-sample with a horse. And company president Aaron Clayton took his year-old

Labrador, Tucker, for a swim in the Charles River outside our offices and began teaching him signals for "swim upstream/swim downstream" successfully. Upstream is harder, but Tucker didn't mind; he said it was a lot of fun.

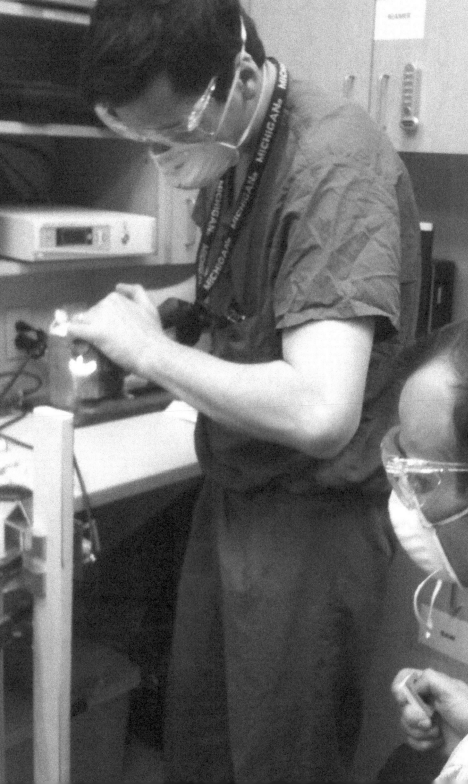

7

Expanding the Conversation

From aquariums to zoos to shelters to dog training classes to specialized dog training, clicker training has made its mark. More recently, it is rapidly gaining a foothold as a valuable addition to human training programs in all sorts of arenas: teaching specific gymnastics skills, training safety aboard fishing vessels, bringing calm to lives of autistic children, giving pet owners basic clicker skills. The list is long and surprising. Here are some examples.

And, Karen reminds us, it all begins with something fundamental that we share with the rest of the animals: a nervous system that's designed to learn, and learn fast, from little whispers of good news around us: a scent, a glimpse, a rustle in the grass—or a click.

These orthopedic surgeons in training are taking turns using TAGteaching to hone precision cutting skills (here practicing on wood). Photo: Karen Pryor

Positively Healthy!

One thing clicker training gives us is a new level of trust and communication—something you may have noticed with your pets. For Karen, it's particularly illuminating and moving to experience this with captive animals that could never be pets; it makes their entire care much more humane.

Imagine you are with me on a backstage visit to a big US zoo. One of the keepers offers to show me the husbandry training (training for medical care) that they're giving the baboons. Carrying a small kit of equipment, she sits down at the bars of an enclosure containing perhaps 20 baboons. She calls over a young female by name, puts a stethoscope through the bars, and listens to her heart and lungs. She signs "open" with one hand (fingers to thumb and then spread apart) and inspects the baboon's open mouth and teeth. Then she picks up a needle-less syringe to practice giving a shot. The baboon, seeing the syringe, presses herself sideways against the bars. The keeper starts to touch the syringe to the animal's hip. The baboon immediately backs up along the bars. The keeper tries for the hip once more. The baboon, looking cross, backs up again, bringing her shoulder level with the syringe. The keeper laughs, and says to me, "Oh, wait, I see. I've never actually worked with her before, and she's explaining to me that she gets her shots in her shoulder, not her hip!" Click.

Shape for health

Oh, but don't shots hurt? Surely that's not positive reinforcement! Doesn't the baboon object to being poked with needles? She would indeed if you just walked up and poked her, but you shape the behavior instead.

May I touch you with a pencil point? Yes? Click/treat. How about a needle-less syringe? Yes. Click/treat. How about a prick with a tiny needle? May I slip the needle in? Pretty soon immunizations and antibiotics and even blood draws are a snap. In fact, it is easier to accustom animals to the prick of a needle than to the weird smell of the alcohol used to disinfect the shot site. That takes some tact.

Did you know?

In case you're wondering, you draw blood from a baboon from the arm, inside the elbow, just as with a person. Tigers and lions have a big vein running down the top of the tail; they learn to stick the tail out through the bars for blood draws. Rhinos present the front leg; elephants give you an ear. With most hoofed stock—okapis, wildebeest, and so on, you use the jugular vein, in the throat. Routine blood draws can identify every-thing from pregnancy to pinworms, but drawing blood from every animal in the zoo even just once a year was incredibly traumatic and logistically impossible using physical or chemical restraint. Today, positive reinforce-ment training keeps zoo animals healthy and happy in a whole new way; and the familiar keepers, not the scary veterinarians, can do the husbandry tasks themselves.

Zoos Then and Now

Once marine mammal trainers discovered the efficacy of operant
conditioning, one of the next populations to adopt the practices
was the zoo community, and what a difference it has made.

When I was little, most zoos consisted of a series of cement-floored, iron-barred cages with a different kind of animal in each cage. One zoo director told me that in those days he hired as zookeepers "…anyone who could see lightning and hear thunder." They had to be men, too; the main jobs of a zookeeper were to shovel food in and droppings out, and to handle heavy loads such as bales of hay.

Now it's completely different. Zoos have evolved into scientific and conservation organizations. Many species are kept in "natural" habitats that provide at least some of the elements, like nest holes and mud wallows, that would make those animals comfortable in the wild. Zoos keep elaborate international breeding records so individuals of a given species can be tracked and swapped around to maintain genetic variability in the captive population. Curators have PhDs and accomplish valuable research. Keepers are often biologists, ecologists, or animal behaviorists. Lots of them are female. And here's another big difference: more and more keepers are operant trainers, too.

Many zoos are now using operant conditioning to improve the wellbeing of their animals. Targeting—nosing and then following a pointer or other object—enables keepers to move animals around without frightening them. Clicker training provides mental and physical stimulation, enriches the animals' lives, and can even save lives. Once upon a time zoo animals had to be immobilized with a dart gun to get medical treatment. There's

some risk involved—it's hard to judge the dosage—and the animals univer-
sally hate the experience and often exhibit extreme stress, which, of course,
skews the results of any blood tests you might want to take. Now, clicker
training enables zoo vets and keepers to weigh the animals regularly and
to perform annual physicals, vaccinations, pregnancy exams, blood draws,
hoof care, and treatment for illness, with the calm cooperation of their
patients.

The last time I visited, the Dallas Zoo was a good example of a zoo
in transition, from the old view that you should not interfere with the
animals in any way, to an awareness of the clicker technology and its usefulness. California zoo-training experts Tim Desmond and Gale Laule, of
Active Environments, Inc., started an operant program with the Dallas
Zoo elephants. Gradually the same positive training techniques began
spreading through the zoo, with keepers learning on their own and helping
each other.

Val Beardsley and Bonnie Hendrickson invited me into the chimp
house. With clicks, praise, verbal cues, and treats, Val showed me how she
can take temps with an ear sensor, file nails, inspect teeth ("open wide!"),
ask the chimps to present a shoulder or hip for injections, and position
their hands, feet, and bodies wherever she needs them to be. Besides the
treats, it is obvious fun for the chimps and they politely wait their turn to
play doctor.

We hopped in the golf cart and went behind the scenes to the holding
barn for Tut, a caracal. A caracal turns out to be an African desert cat of
about 30 pounds (think cocker spaniel-size), sand-covered all over, with the
most amazing black and white tufted ears. Keeper Kristin Streebel showed
me Tut's daily training routine. He works for verbal cues, clicks, and meat
treats (passed through the cage mesh on a stick—he's quick and he bites).
Tut turned now and then to hiss at me and my video camera, but he never
quit working. Great clicker cat.

Here are the cues Tut knows and responds to: sit, down, spin, high paws, come, get on the scales, and "shift" (meaning dash through the door into the next enclosure). Tut spends his nights in the barn and his days outside where the monorail passengers can see him. So he needs to shift reliably, and he does. At about 1000 miles an hour. For a click and a treat.

Then Megan Lumpkin showed off several of the zoo's okapi herd. I have always loved okapis, a rare animal now (Ituri Forest Congo cousin of the giraffe) about the size of a horse, with big dark eyes and huge furry ears and long, long tongues. A young female came into the barn and put her nose on a target (a piece of pipe) held by one keeper, while another keeper touched her back and sides and gave a simulated injection in the haunch.

We visited a mother okapi and calf outdoors in the sunshine. Mama showed that she could stand patiently to have her hooves cleaned and cared for. Baby showed that he could buck, run, jump, kick, lick at my camera lens, and try to sneak out an open gate. Like domestic horses, all hooved stock in zoos tend to need regular hoof trimming and cleaning since their hooves don't wear down the way they would in the wild. Only in the last decade, with the rise of simple clicker technology that keepers can learn by themselves, are these hooved animals beginning to get that care.

Keepers don't often get time to visit other parts of the zoo and see other training in action, so throughout the day I had two or three zoo staff members with me, grabbing the chance to watch other trainers in action. Our next stop was the elephant barn. Traditionally, as with horses, trainers have managed elephants by pushing them around physically. They used to control the elephants largely with positive and negative punishment, negative reinforcement, and social dominance. It takes skill and practice, and it can be very dangerous; when elephants get mad, keepers can get killed. Clicker training has made it possible to care for elephants through a partition, with clicker and treats, a system called "protected contact." It is easier on everyone.

Keeper Audra Cooke showed us how Jenny, a beautiful African elephant, targets and stations for injections and other care. She presents each foot, through a gap in the bars, for foot care and trimming; and she can flare her magnificent ears on the cue, "ears!" She lost her cool, once, and squealed and whirled away from us, tramping her front feet nervously. (Audrey thought she was nervous about exposing her hind feet with several strangers in the barn.) In the old days, that would have provoked a stern response from the keeper. Now Audrey just waited a few seconds, safely on her side of the partition, until Jenny calmed down and came back for more clicks.

There was lots more. The warthogs, homely but charming, targeted and accepted lots of body contact, from simulated injections to petting. Mrs. Warthog seemed to like having her foot-long head hair combed and played with, and Mr. Warthog actually lay down to better enjoy a belly rub.

The rhinos are all target trained, foot-care tame, and willing to have any part of themselves, including inside the mouth, inspected and treated. I borrowed a target and trained one of the rhinos, and scratched it behind the ears, which it accepted (along with some banana segments) with a pleasant look.

Best of all, though, were the tigers. Jay Pratte, the tiger keeper who had organized this whole busman's holiday, took me to the tiger barn—a Maximum Security, Triple Door, Lock-down type residence—where he and Sherri Falls put two male Sumatran tigers through their paces. Shift. Come. Down. Swing your tail under the bars. Scootch forward so I can reach your tail. Stay, while I sterilize your tail with alcohol and give you a pretend—or real—injection. Back up. Click. Thank you, you may go.

The tigers *love* this and rush to their training stations, grring and rumbling, but, like the rhinos, with pleasant faces. Their favorite behavior—and mine—seemed to be "up." The keeper stands with his or her hands as high on the mesh as they will go. And the tiger, in a lightning move, stands in a matching pose, with front paws over his head on the mesh, on the

234

other side. Jay Pratte is a tall man. But the tiger, on its hind legs, is a lot taller. An impressive creature to have for a friend.

As in many zoos today, keepers in different departments are at different levels of skill in their training. But everyone is trying to learn more; even birds and reptiles get positive reinforcement for cooperating in medical care and general husbandry and handling. It's a transformation for the zoos, a fascinating new way to interact with the animals for the keepers, and, in some cases, a life-saving, or even a species-saving, technology. Click!

Fun with Rats

*What better place to apply clicker principles than to the learning
and lives of lab rats?*

When I lived in the Pacific Northwest, I kept a brown and white pet rat named Lucy. She would come when called, so I could give her the run of the office. She was a delightful pet, cozy, chatty, and full of enjoyment of life. When I moved to Boston, I re-homed her with a neighbor's little boy, and he loved her, too.

Lab rats

In the course of writing *Reaching the Animal Mind,* I interviewed some leading neuroscientists and met some of their rats. Thanks to my history with Lucy, I now took a real interest in those rats. Most of the experimental rats were fitted out with cranial implants—metal hats—that allow direct study of the brain in action. This headgear didn't seem to bother the rats. What bothered me, though, was an absence of operant training in some experimental designs. Some scientists were asking for straightforward tasks from an operant standpoint—press this if you see that—but it seemed to me that they didn't explain the job to the rat in a way the rat could understand.

One day here in Boston, I received an e-mail from a Harvard neuroscientist named Bence Ölveczky. The students in his laboratory were reading *Don't Shoot the Dog* and wanted to find out more about using operant conditioning with their rats. As I understood it, the work they were focusing on was this: what happens in the brain when the animal learns complex behaviors, such as a behavior chain. Isn't that right up our alley? Yes it is. Would I come over and talk to them? Yes indeed.

I went to the lab, which is about 15 minutes from my house, and met with Dr. Ölveczky and some of the students. I gave a talk about clicker training to a group from several departments. And I saw some rats.

Enrichment

Because of the cranial implants, the rats have to live alone; you can't have another rat nibbling on that fancy headgear. Each rat lived in a transparent box with deep, clean, suitable litter. The students asked me if I knew ways to make their rats even more comfortable.

I consulted by e-mail with Virginia Broitman, a clicker expert and rat lover. Virginia provided a list of what zoo trainers call enrichment items for rats: things to chew and manipulate; periods of darkness; chances to explore; novel foods. I forwarded the list to the lab, along with a memorandum on the requirements for a good marker signal and some thoughts on reinforcers.

What kinds of behaviors are needed?

Examining and probing the rat's brain through a cranial implant is a painless procedure, but the brain needs to be held still during the process. Could a rat learn to hold still voluntarily, instead of being under total physical restraint? Probably, I said. People are doing it with monkeys through clicker training.

A month later, I had lunch off-campus with Adam Kampff, a post-doc in the Ölveczky lab, and his wife Eva, another neuroscience graduate student. Adam told me that a student had successfully trained one of her rats to put its head in the right place and keep it there, but when she lowered the clamps on the headpiece prior to actually activating the probes, the rat panicked and pulled away.

"She's just moving a little too fast," I suggested. "It might help to break the training down into smaller steps."

Does that help? We'll see.

Another visit to the lab

In late February I made another visit. The lab had expanded into big new quarters. The rats were arranged so they could see each other. Guess what else? All the rats now had pieces of cardboard or paper and wood to chew on and manipulate.

Now everyone was training in several short sessions daily, rather than letting the automated training run for hours. Good! And the student working on voluntary self-positioning had trained three rats to go into position and hold their heads still on cue, for longer and longer times. She showed me that. Good job!

Another student had developed a clicker trained, stress-free way to move rats around. She placed one end of a foot-long, flexible piece of thick tubing (like dryer vent tubing) in the rat's home space. The rat ran into the tube. The student carried the tube to the experimental chamber, and the rat ran out the other end. A happy, calm rat and a happy, calm investigator.

The student also showed me her rat hotel. She had glued two cardboard boxes on top of each other, with holes in between, rope ladders, and toys— for ratty, after-work fun time. A terrific non-food reinforcer!

Grad student Raj Podar wanted to use the computer to teach rats the basics of clicker training. The idea was to automatically teach rats to understand working for the click, to learn specific behaviors, and to learn that there are cues for each behavior. These pre-trained rats might have a real advantage in learning new tasks for subsequent experiments.

Raj switched his rats from day to night, because that's their natural alert time. While we could watch them on the computer, the rats themselves worked in darkness. A simple kindness, yes, but also biologically appropriate and, maybe, helpful to the learning.

We watched on the computer monitor as four rats had their training sessions. Three were learning well, pressing the left stick on one cue, the right stick on another. The fourth rat, however, had gotten stuck on one side, and was in a classic extinction curve, giving up all behavior, curled up in a corner in despair. Together, we all managed to figure out what may have happened and came up with a way to tweak the cue presentation method a little to see if that helped. Fun for me!

The training game

After lunch we played the training game. Students clicked each other for turning in a circle, putting paper in the trash, picking up a rock from the bamboo planter in the window. Laughter and applause. A first shaping experience.

Then I demonstrated shaping with a rat. The first rat we moved to a work table was vastly more interested in climbing out of her transparent box and exploring the tabletop than in food or work. We tried a second, more timid, rat. He stayed in his box gladly, and proved to be willing to work. I captured the behavior of sitting up on his hindquarters several times. Soon he was obviously doing it intentionally and holding the pose longer and longer. I suggested that someone else now tackle the shaping task of putting the behavior on cue. We put the rat back on the rat shelf so he wouldn't go on offering the behavior in vain.

The students watched in silence. I was not dismayed by the lack of verbal behavior. I thought they were probably engaged in a behavior they were highly trained for: focusing, thinking, and learning. A lot. And fast.

And now...

This essay was written in 2010. Some of Karen's students have kept in touch with her. They are training creative rats with clickers. They are carrying the technology to other laboratories where they are now scientists. They are teaching positive reinforcement training to students of their own. Karen has visited two such labs. At Fundacion Champalimaud, in Lisbon, Portugal, where one of her Harvard protégés now has his own lab, newcomers play the training game, and experiments are built around the clicker. In Arizona, Russian scientist

Irina Beloozerova and her colleagues study movement in the nervous system by training cats to negotiate obstacle courses—with clickers and minced chicken. And in both of these laboratories the text book for new trainees is not some technical tome, and not even Karen's popular book Don't Shoot the Dog, *but, to her surprise, her earliest (and simplest) book about training, the story of her time as a dolphin trainer,* Lads Before the Wind.

How to Write a Scientific Paper

Part of expanding the conversation about clicker training is sharing and
communicating the research behind it. Clicker training is in an odd position
where the practitioners have adopted innovations at such a rate that they
have outpaced the scientists who test the principles and accept or reject
the reasoning behind them. Science is now playing catch-up to practice and
Karen is very much at the forefront of this effort. Not only has she tried to
integrate the fields of ethology and behaviorism; now she's spanning the gap
between popular knowledge and scientific validation. It's been an
uphill battle to make clicker training scientifically "respectable."
Here, then, is a view from a different bridge.

New information—however interesting, amusing, or useful—is not
accepted by scientists until it has been published. This does not mean
published in the *New York Times* or in a best-selling book; it means
published in a peer-reviewed journal. A peer-reviewed journal is run by a
scientist-editor or editors. The editors look at your manuscript and then
send it out to two or more "peers," that is, people who work in the same
field and are well known (i.e., well-published) themselves. These reviewers
remain completely anonymous; presumably you will never know who they
are. They critique the paper and often suggest changes. You meet the crit-
icisms and revise the paper. Eventually the paper is either published or
rejected. If it's rejected, you can chuck it in the wastebasket or send it to
some other journal, whereupon the process begins again.

One (long) story

I've published a number of papers in the peer-reviewed literature over my lifetime. The best-known paper was the result of an event at Sea Life Park in the 1960s. Ingrid Kang Shallenberger and I accidentally developed a dolphin that could think up her own tricks and, in fact, was spectacularly good at it. The US Navy was interested and funded a repeat of the process with a naïve dolphin, requiring us to collect data throughout. The resulting paper (the "creative porpoise" paper: see "Find Out More!" page 261, for this and other papers referenced here) has been cited in other scientific literature more than 600 times.

But for me, that was that. No fan mail. No feedback whatsoever, except that once in a while some professor will tell me that he assigns the paper to his students. In spite of all the citations over the years, people seem to regard this study as a quaint, one-time event. The study is used to show that dolphins are really, really smart, or that applied operant conditioning can occur in a way that students will not find boring. The important aspect to me—that *many* animals might be capable of innovative thinking, given the right training situation—did not catch people's attention.

What the scientists ignored, however, the trainers picked up on. With my book on positive reinforcement, *Don't Shoot the Dog!*, the advent of the clicker in 1992, and the new ease of communication due to the Internet, creativity training spread fast in the non-scientific world. Many people began teaching animals other than dolphins to be creative. Dogs. Gorillas. Sea lions. Birds. Even fish.

One day my friend, colleague, and mentor Sheila Chase, a professor at Hunter College in New York, sent me a Call for Papers from the *International Journal of Comparative Psychology*. They were putting together a Special Issue on behavioral variability. Sheila suggested that I submit something, bringing the "creative porpoise" paper up to date by describing the growth of training for creativity that had developed since then. So I wrote to the editors of the journal and said I'd like to offer an update. They wrote back, very surprised (I think they were surprised to learn I was still alive), and said, yes, they'd love that. And with my well-published friend Sheila Chase as co-author, we put a paper together and sent it in.

The reviewers objected vigorously. One didn't understand the paper, period. Another said the paper was too casual, that it read "like a magazine article." Okay, I can see that; we don't need colorful writing, and, in fact, being too informal can actually get in the way of comprehension. We did a careful revision and sent back the paper. The editor objected to *lots* of things in the revision. We did another thorough revision, taking a different approach. He was even more upset.

Among other things, the editor wanted references for every "statement of truth" we made about modern training. I muttered to myself, "Yeah, but there *are* no references. Because there is nothing in the literature. That's why we're *writing* this paper!" Back and forth went the manuscript between me and Sheila, with Sheila crafting key paragraphs and digging up an amazing number of references that did support what we already knew. (One of my favorites is Grice's 1948 study demonstrating that even the briefest delay in the marker can mean the experimental animal learns nothing.)

Anyway, we kept working. Between November of 2012, when we drew up the first outline, and May of 2014, when the paper was in press (that is, set in type and being printed in the journal), my files show that we created 18 different iterations of the manuscript. Do you want to read the final version of "Training for Variable and Innovative Behavior?" You can find it online at http://escholarship.org/uc/item/9cs2q3nr.

If you don't want to plow through the whole paper, below is the abstract. It says very much what we set out to convey in the beginning. It just took a lot of work—and a lot of skilled help from the editor, David Stahlman—to actually get there.

Abstract

This paper provides a summary of a 1969 report of the spontaneous emergence of innovative behavior of a dolphin, a replication of this event through training in another dolphin, and the effect this work has had on current animal training technology. This paper provides a review of laboratory-based research in support of some of the procedures found effective in modern animal training in developing innovative behavior, specifically use of the conditioned reinforcer to mark a behavior, differential reinforcement of variability, and intentional use of positive reinforcement procedures. The authors describe specific processes for establishing innovative skills, practical applications presently in use with animals, consequent human and animal welfare benefits, and suggestions for further research.

More papers

I'm not done. This year I published not one but three papers in the scientific literature. By chance they all came out in May! One was this creativity paper. Another reviews my history as a marine mammal scientist. The third is a long paper on modern animal training, co-authored with Ken Ramirez, which has been published as a chapter in a textbook on classical and operant conditioning.

I'm now working on two other chapters in scientific texts, one on creativity in animals and another on my history with behavior analysis. I'm also working on more papers with Sheila. I'm currently co-investigator on a two-year research project involving marker-based training of humans. I'm also on the thesis committees of several Hunter College graduate students. I hope that new papers may come out of those projects as well, demonstrating more things that every good clicker trainer knows, but that are "not in the literature." Yet.

Traditional Obedience vs. Clicker Classes for Beginners

How does a beginner clicker class compare with a traditional beginner obedience class?

To most people, a beginner obedience class is a class where the dogs learn to sit, down, stay, heel, and maybe come. Five behaviors in six or eight weeks. Is the beginner clicker class an ordinary six-week-long basic pet owners' obedience class, with clickers added?

I don't think so.

The goals are entirely different

ClickerExpo faculty member Kathy Sdao has given a lot of thought to this question. Clicker training, Kathy says, is a way of communicating with an animal, using a universal language that animals understand from the start. "Control of the animal's behaviors then flows as a byproduct" of that communication. (As do a lot of other things, such as the animal's much improved ability to communicate with you.)

Kathy goes on to say: "Behavioral control, however, is the principal goal of standard training. Communicating with the animal is the means to this end."

This is a big, fundamental difference. Most regular dog classes reflect that behavioral control goal. Control behavior: stop behaviors A, B, and C and be able to command and get behaviors X, Y, and Z. Period. Communicate? Sure—with leash, gesture, voice: but just to control.

In a clicker class, we should be teaching people to use the tools of communication that this methodology provides. If the people know the method and learn how to click, the dogs will come along just fine. Furthermore, the humans can get all the behaviors they want, not by recipe but by communication tools.

But some dog-training instructors believe that pet owners don't really want to learn to click or communicate with their dogs; they just want to get to the point where the dog will sit because they say "sit," and clicks and treats are no longer needed. I think that perhaps this problem arises when—and because—some pet classes with clickers still follow the same schedule that pet classes with choke chains used to: six weeks to learn sit, down, come, stay, and heel. What more does the "pet owner" need or want, anyway, right?

Sit, down, and stay, however, are not good clicker behaviors. They all involve stopping the dog from moving. The aim is to remove behavior, not to build it. There's nothing to click, except longer and longer periods of doing nothing! This is not what clicker training is for, and it doesn't teach the dogs how to think; it doesn't make them into clicker dogs. So maybe for a start, in our perfect beginner class, we need different behaviors: walking, moving, active behaviors.

We need a whole new plan

But it's not just that the goals of the two types of classes are contradictory. The first big complaint I always hear about introducing the clicker into a basic obedience class is, "Oh, but beginners are too clumsy; they can't handle the leash, the clicker, the treat, and the dog all at once!" I think we need a whole new plan. Maybe a clicker class should teach the skills the *people* need, not the skills the *dogs* need: How to observe. How to "catch" an incredibly brief movement with your clicker. How to hand over 10 treats in 20 seconds. How to choose what to click next. How to click when you have a leash in one hand and treats in the other. How to focus on your dog, not on what people must be thinking about you. How to develop a cue. How to use cues that work, not cues that don't work. How to read your dog's mind.

For instance, Massachusetts clicker instructor Tibby Chase told me about her new way of teaching people to teach dogs to "walk by me." It involves lots of clicks, lots of treats, lots of traffic cones—and no leashes. Leslie Nelson of Tails-U-Win takes polite walking a few steps further—75 steps, in fact—and her students are conquering competition heeling. In both cases, it's the people who have to learn; the clicker-wise dogs figure out their job from the start. Maybe we need to be clicking the people, not just the dogs!

Building the Perfect Clicker Class

The "perfect" clicker class is still evolving, but here are some of
the concepts that work. What would you add?

Clicker training is not about physically manipulating a dog, although that takes a different sort of expertise. Instead, it requires acquiring both mechanical skills (handling the leash, dog, clicker, treats; timing; observing clickable moments) and conceptual understanding of the process (shaping, adding cues, and so on). Isn't it overwhelming to the pet owner? Aren't they all thumbs?

Break it down

Right. So our job is to clicker train those skills. Help them practice. ClickerExpo faculty member Kathy Sdao has devised some wonderful exercises for increasing the necessary manual dexterity to click and treat rapidly, such as giving each student a cup full of dried beans, and an empty cup. Picking up one bean at a time, and dropping it into the empty cup, how many beans can you transfer in 10 seconds? Go! Manual dexterity and developing a practiced fluent movement are the goals here. And everyone gets reinforced, usually with Hershey's Kisses.

Another tactic is to divide the group into pairs to work on targeting. You tether or crate one dog (or have a bystander hold its leash) and ask both people to work with the other dog. The owner of the chosen dog holds the leash and treat supply, and gives a treat after each click. The other person clicks for some simple behavior—touching a target is a good place to start.

One person has to think only about holding the target, getting the behavior, and clicking it. The other person concentrates on listening for clicks and giving the treat promptly (and even that can be hard at first!).

After two or three minutes—the teacher can call time—they change places: now the owner is clicking and the partner is treating. Both are gaining skills, both are learning to focus without being swamped and confused, and the dog is learning that total strangers can click and treat and make sense: good news that helps the dog to calm down.

Then they put that dog away and do the same thing with the other dog. Even in the very first lesson you can give them additional behaviors, or start leading the dogs in circles and short lines with the target. You can swap pairs around so that everyone experiences three or four dogs and two or more partners during the first class.

This is what TAGteachers call "BID," or Breaking it Down. What other ways can you think of to break down the mechanics of clicker training so everyone can get good at it in 10 minutes?

Beginner people need a high rate of reinforcement, too

Now, think about this. We advocate a high rate of reinforcement for beginner dogs; what about beginner people? We should be using clicker-training principles to teach people, too. When I'm demonstrating in class I mark people while they are clicking their dogs successfully. I might mark a specific stated point ("Click when the dog's nose approaches the target") or for improvement as they work on the task. Or, I might just mark people by surprise—for attitude, for laughing, for helping someone else, for any behavior I see that I like. Do you give the people treats? I might, at first, but soon just the feeling of success is *plenty* of reward for adult humans. That's TAGteaching; it's a great way to shape behavior in people. (There's lots more useful information on www.tagteach.com.)

Throw the traditional curriculum out!

What about the behaviors the dogs should learn? At the very beginning, it's important that the dog discovers that what he's doing is what made the owner click. If what he is doing is sitting still, and he was sitting before, during, and after he heard the click, how can he tell exactly what he got paid for? When training heel, how much easier is it if the dog is moving more or less alongside his owner and gets clicked every third step? His muscles learn, his muscles can repeat the behavior, and, in no time, he is "training" his owner to click, by walking close to her left side as she is moving, too.

But for beginners, forget sit, stay, and down—all these controlling, movement-inhibiting behaviors. They're hard to learn and hard to teach. Start with getting the dog to target to the owner's fist, and then move to touch the fist, or target back and forth between two people. Or, follow a target stick, ending by following it under a chair or in and out of a box.

How about shaping? Shaping—without luring, leash-leading, or pushing the dog around—is the key skill for building communication with the clicker. Once the person and the dog understand that game, they can learn *any* behavior together. Everything stems from shaping. "Oh, that's much too hard for beginners." No, they learn it just fine. It's much too hard to teach only if you are used to teaching by verbal instruction, because shaping is a real-time, non-verbal process, like dancing. Explanations aren't much good.

So set the students some easy capturing task—capture a head dip, a head turn, a paw move—and let them shape it into a bigger move—and mark their successes as they click and treat their dogs. I've seen Virginia Broitman and Sherri Lippman teach a whole roomful of dog/person couples to capture, shape, and put on cue a paw gesture (high-five, say) in 10 minutes or so. And everyone learned.

What about continuity from class to class?

Who says every class has to build on the class that went before? That's a given in a standard class; in the first class or two you work on a short sit, a vague sort of heel, maybe a forced down, an on-leash come. In the second week, you extend the behaviors. And so on and so on. Forget it!

Emma Parsons devised a six-week beginners' clicker class for Tufts University Veterinary School in which the exercises each week were completely different. The learning goals were the same—and the goals were for the people: learn to use the clicker to mark behavior; learn to reinforce at a high rate; learn to observe and capture small moves or brief attempts and shape larger responses from them. Learn to let the dog work it out rather than trying to shove him through it each time; learn to leave behind any urge you might have to correct mistakes, and instead, catch and click success. Students often worked in pairs, training one of their dogs at a time.

The first week, they did some basic on-leash stuff. The next week was "a visit to the veterinarian," with examination tables, students taking turns playing the vet, and dogs moving from one station to the next, getting clicked and treated for pretend ear exams, for being lifted on and off the table, for pretend nail trims, and so on. The third week, everybody tried some agility obstacles, set very low and easy, just for fun. The fourth week was aggression week: exercises for meeting, approaching, and passing other dogs politely. Tricks. Tracking. Trading dogs. Every week was different.

What happened?

One man dropped out: he did not want to train other people's dogs, or let them train his. The rest of the students came more and more reliably, often showing up early and staying late. In six weeks the students were all skilled and versatile with their clickers. The dogs were all focused, enthusiastic, and attentive. "Bad" behaviors such as jumping, bolting to the end of the leash, sniffing the ground, and concentrating on the other dogs had disappeared. People never knew what to expect next, and, therefore, never suffered the guilt of coming unprepared because there was no homework. No one noticed they were "practicing" when they clicked their dog at home for some good move. They came to class with surprising new tricks or skills to show off, every week. No one was complaining about their dog's "problems." Everyone was bragging about his or her dog, and rightly so.

Now they loved and understood their dogs more. I recall seeing one woman in tears at the end of her six weeks, because she was so grateful to have come to know her nice dog, a dog she had previously regarded as a nuisance and a burden and had been thinking of giving away.

And, just in case you think, "Oh well, vet students, they were already pretty dog-savvy," they were just like any other beginner class. While some of the dogs were pets, some were not. With permission from the vet school, many of the students had been allowed to use laboratory beagles—adult animals raised for research. They had never been out of a cage in their lives. I recall that one of them, when first carried out to the training field and placed on the grass, was so overwhelmed that he fell over.

At the end of six weeks, all those beagles were normal, and all were clicker-trained. So were their people.

A better way?

Isn't this rather different from drilling five behaviors over and over at home (if pet owners even bother) and in class? If the owner and dog have tried out clicking together for 20 or 30 different behaviors, a single behavior is no longer such an issue, and the dog is more likely to do it willingly and attentively anyway, having a richer reinforcement history in general.

Did this suit everyone? Of course not. It is tempting to give a lot of attention to the Unhappy people—the "Yes, but" folks; the complainers who want sympathy, not solutions; the arguers who want to prove you wrong; the resisters. It is more clickerish, however, to reinforce the ones who are trying to learn instead of giving attention to behavior you don't want. Let the resisters watch and learn, or let them leave. The successful learners will keep your classes filled.

Karen wrote these thoughtful pieces in the period from 2003 to 2006. Together with Aaron Clayton, president of Karen Pryor Clicker Training, plans were already underway to translate these ideas into online courses, programs, and workshops designed to teach people to be better clicker trainers and better clicker training instructors, too. Today Karen Pryor Academy has online and on-the-ground teaching programs at dozens of locations in North America and worldwide. Thousands of people from around the globe have completed Karen Pryor Academy courses.

Teaching People Teaching Dogs

Karen's first experiment with "clicking" humans taught her as much as it did her students.

In 2005, company president Aaron Clayton and I, with ClickerExpo faculty members Virginia Broitman and Sherri Lippman, traveled to the West Coast to take on an exciting assignment. At the request of one of our business clients, we had developed a specific curriculum for teaching pet owners the basics of clicker training. We now had the job of teaching six business executives enough about clicker training to enable them to begin to teach other trainers in their employ.

Our six class members were experienced "lure-and-reward" dog trainers. Some were also experienced in correction-based training. Most had been exposed to some clicker training and had used it but had never really adopted it. Our goal was to enable these six trainers to understand enough about clicker training to be able to teach others to use the curriculum. And we had two days.

Day one: we begin

The first task was to teach them clicker timing and mechanical skills. We started with games and exercises developed by TAGteachers (and fellow ClickerExpo presenters) Theresa McKeon, Beth Wheeler, and Joan Orr, to teach children to use the clicker while learning athletic skills. We then went through a series of other games designed to introduce clicker training skills one at a time. We used carefully crafted positive language, also based on TAGteach practices. We reinforced with attention and recognition as well as with verbal praise.

We led our students through examples of the kinds of exercises and games that made up the proposed curriculum. We had them practice teaching elements of the exercises on each other. We led the group through team-teaching exercises in which they clicked each other. In the afternoon, we brought in some dogs from the San Diego Humane Society so everyone could practice target training and free-shaping with friendly but naïve animals.

Reinforcing humans: using my pinger toy

In addition, I used a toy I found that made a "ping" sound (substantially different from the click) throughout the day to mark people's behavior. When someone said something insightful, or clicked a partner's moves especially well, or burst into laughter at a dog's or human's success, I "pinged," and handed him or her a Hershey's chocolate. I was reinforcing for participation, for signs of enjoyment or enthusiasm such as laughter, and for improvement in any exercise, especially if someone had been having difficulty. I tried to make sure I found some reason to ping each person at least two or three times in the course of the day. I didn't explain myself, and I didn't verbally identify what I had pinged; I just did it.

At first the ping, and especially the handing over of a candy, seemed to be mainly an interruption. "Huh? What's this? I don't want a candy!" By the second or third personal ping, however, people glanced at the candy, took it without comment, and stuffed it somewhere: pocket, bait bag, nearby purse. No one questioned it or remarked on it; they ignored it, at least verbally. The fact that they were accepting and even collecting the candies, however, suggested to me that the ping sound was becoming a significant reinforcer.

Of course I was also, unavoidably, pairing the ping not just with candy, but with visible public attention from the head trainer, me—and that is not necessarily rewarding! So I was careful not to identify specific behaviors verbally, which might make the students focus on "pleasing" me or avoiding me, rather than just doing their stuff. And I was careful not to embarrass or distract them as I handed over the chocolate by offering verbal praise (see "The Perils of Praise," page 166), or social interaction such as smiles or eye contact, which would oblige them to give a social response. I hoped my public attention would thus be at least a neutral experience rather than stressful. Mainly I was hoping that the class members might come to experience for themselves the profoundly elating sensation of earning an unexpected reinforcer.

I also carefully watched my timing when I pinged. I always pinged while the behavior I liked was still going on—*never* as praise after the behavior was done. If you use the sound after the behavior is over: a) it is useless, since your learner is now doing something else, and b) it immediately acquires some of the negative values of praise. It draws unwelcome attention from others; being meaningless it seems insincere and therefore manipulative; and it probably seems a little condescending. With or without interpretation you can see that it is now unwelcome (aversive) by the look on the person's face. A ping during the action? They absorb internally and keep going. A ping when the action is over? They give you the fish eye.

I had never done exactly this in a group before. Was it working? Reinforcement is defined by its effect on subsequent behavior. Time would tell.

Taking the pinger into the trenches: Day two

The very next morning, before class began, the KPCT folks and the six members of our class were working together to set up the room. Everybody seemed very upbeat and excited. Sherri said something funny, and I pinged her. Instantly the six class members' heads swiveled around.

"Who got pinged?" "What did she do?"

My heart leapt. *Yes!* It's working! The ping had become an unequivocal positive reinforcer, doing its job for them and for us, in a very, very challenging situation.

In the afternoon, we brought in eight pet owners with their own dogs to be a "practice" class. None of them had any clicker training background. Many of them had no training at all. There were puppies. There were clueless, bouncy, adolescent dogs. There were at least two very reactive dogs, one of them plunging, snapping, and snarling at the other dogs. Whew!

We gave each owner a specific chair to sit in, so we could spread the dogs widely around the room. We put visual barriers—more chairs—between the most excitable dogs and their immediate neighbors. We passed out clickers, treat bags, and treats. Sherri took the really aggressive dog and its owner into a quiet corner and worked with them exclusively. Virginia and I started teaching and demonstrating exercises. Our six executive trainers took turns leading the pet owners through games and exercises from the curriculum; they did it flawlessly.

Then our executive class members each chose an owner and dog, drew up a chair, sat down, and started teaching the owner how to build the behaviors that were being demonstrated, click by click. I went around pinging and giving candies to the teachers and the newbie dog owners, too. I was catching behaviors I liked: laughter, intense concentration, rapid reinforcement. Since I knew the ping was working, those behaviors would be bound to increase.

By the end of the class (90 minutes long!) every dog was sitting or lying attentively at its owner's feet, gaze *glued* on the owner's face, a big, new experience for some owners. Every dog had a good start on two or three tricks. Many had considerably improved their manners, including skills such as taking the food gently and maintaining self-control and attentiveness.

Even Sherri's student, the owner with the aggressive dog, had made huge progress. The dog had learned "sit," "leave it" with eye contact, and a couple of tricks. It had stopped creating an uproar at any glimpse of other dogs. Sheri excused them a little early, asking us all to step back against the walls and give them room. They passed quietly through the room and out

257

the door, the dog being clicked and treated for looking at the owner, not at the other dogs, as they went. The whole place—owners, trainers, and we KPCT folk, too—burst into applause.

The head honcho came in to see how his trainers were doing. They were all doing great. They were a success. What had begun as barky chaos ended as complete serenity, with radiant smiles from teachers, from pet owners—and, judging by the gently waving tails—from the pets.

That's what a clicker class should look like, I thought. Ping!

A Summary Dream

Clickers are not the only topic on Karen's mind.
Karen is a naturalist from childhood, a behavioral biologist by
education, and a gifted observer. On a quiet summer day, Karen
the observer puts herself—and you—
into some other minds.

If God made man in his own image, I figure God made woman in his own image as well, and lots of other things, too. I mean, God might not always care to look like an adult male primate. Sometimes She might want to look beautiful in other ways, and still make other things in her own image. Jellyfish, for instance. What could be more beautiful—and fun to be, too? Jellyfish have nerve centers all around the circle, like lots of little brains that do things together.

"It's cold down here; let's go up to the surface."

"Oh yes, let's!"

Pulse. Pulse. Pulse. Into the sunshine.

The renowned biologist Thomas Huxley famously said that God must especially love beetles, since he made so many of them. I think God especially loves making other Beings. When he made beetles, maybe he didn't tell them "Go forth and multiply," but instead told them, "Go forth and divide." It seems beetles have chromosomes especially good at splitting beetles into new species. Put a species of beetle on an isolated island and in no time you have 18 species of beetles, or 80, big and small, round or long, black, spotted, metallic green, all with different lives.

I think God loves behavior. He made a lot of enjoyable behavior, some of it for beetles. Imagine. You're trundling along the ground like a tank, and suddenly you split your upper armor in half, unfold the tissue-like wings you keep underneath, and fly away. How cool is that!

It's morning outside. A beautiful summer day. Time to go out and watch the teenage squirrels, lately out from under mamma's rules and regulations, chasing each other around in the maple trees. Sometimes they miss a jump and fall to a lower branch, but they're good at catching themselves, and I bet that's fun; it certainly doesn't slow them down any.

How are the summer bugs doing? A huge yellow-and-black tiger swallowtail butterfly, a childhood favorite of mine, flies by, high and fast, as always. A tiny, nearly invisible, thrip lands on my notebook. If I had a magnifying glass, I could show you that this insect is so small it can't afford proper tissue-paper wings. Instead, it has little hairs along a central strut, which are all it needs to get airborne.

Good morning, very small Being. You are the image of your creation, as I am of mine. I can talk and write, but *you* can fly. That's very beautiful, and I bet it's fun, too.

Find Out More!

General:

Pryor, Karen. 1999. *Don't Shoot the Dog! The New Art of Teaching and Training*. Rev. Ed. New York, NY: Simon & Schuster.

Pryor, Karen. 2004. *Lads before the Wind: Diary of a Dolphin Trainer*. Rev. Ed. Waltham, MA: Sunshine Books.

Pryor, Karen. 2004. *On Behavior*. Waltham, MA: Sunshine Books.

Pryor, Karen. 2009. *Reaching the Animal Mind: Clicker Training and What It Teaches Us about All Animals*. New York, NY: Scribner.

For a fine introduction to the science of ethology:
Lorenz, Konrad. 1952. *King Solomon's Ring*. New York, NY: Routledge Classics (2002).

Chapter 1: The SEEKING System

For more on Karen's cichlid fish that she trained to swim through a hoop: *Reaching the Animal Mind*, Chapter 4.

For more on the neuroscience of clicker training and the SEEKING System: *Reaching the Animal Mind*, Chapter 10.

Properly done, clicker training *is* a form of play:
Pryor, Karen. 1981, "Why Porpoise Trainers Are not Dolphin Lovers," Ann. NY Acad. Sci. 364:137-143.

Chapter 2: The View from the Bridge

For solid information on clicker training for horses:
Kurland, Alexandra.1998. *Clicker Training for Your Horse*. Waltham, MA: Sunshine Books.
See www.clickertraining.com/horsetraining

Websites:
The Animal Behavior Society: www.animalbehaviorsociety.org.

The Association for Behavior Analysis: www.abainternational.org.

The Arizona Desert Museum: www.desertmuseum.org.

Bird show developer Steve Martin: www.naturalencounters.com

Shanlung's live journal: shanlung.livejournal.com. For the story about Jackie, the wild-caught adult mynah, see http://shanlung.livejournal.com/135250.html.

For seminars and an on-line course on clicker training horses: www.theclickercenter.com.

For more on Panda and the Panda Project: www.theclickercenter.com/ThePandaProject.html.

For information on the largest and best clicker training organization for guide dogs: www.guidedogs.com

For the main site for information on the many applications of clicker training to human endeavors: www.TAGteach.com

Chapter 3: The Art and Science of the Clicker

For more on the emerging information about the amygdala's role in learning:
Reaching the Animal Mind, Chapter 10.

For more on Kathy Sdao's thoughts on clicker vs. traditional obedience training:
Sdao, Kathy. 2012. *Plenty in Life Is Free: Reflections on Dogs, Training and Finding Grace.* Wenatchee, WA: Dogwise Publishing.

For a study of the efficacy of the clicker compared to the spoken word "yes" in behavior acquisition in dogs:
Lindsay Wood, 2007. *Clicker Bridging Stimulus Efficiency.* Master's Thesis, Hunter College, CUNY New York. By permission of the author this thesis has been made available as a download from http://www.clickertraining.com/node/1960.

For more on micro-shaping for the breed ring:
Pryor, Karen. 2006. *Click to Win: Clicker Training for the Show Ring* (available in a Kindle edition only, 2012).

For more on B.F. Skinner's discovery of shaping:
Peterson, Gail. 2004. "A day of great illumination: B. F. Skinner's discovery of shaping." *Journal of the Experimental Analysis of Behavior* 82, no. 3 (November): 317-28 (www.ncbi.nlm.nih.gov/pmc/articles/PMC1285014/pdf/15693526.pdf).

Chapter 4: Creating a Climate of Abundance

For an example of proper use of a jackpot, see the description of the horse trainer who worked with American Saddlebreds, and for more on the difference between jackpots and non-contingent rewards:
Don't Shoot the Dog, pp. 11–13.

For an excellent explanation of operant conditioning:
Vargas, Julie. 2009. *Behavior Analysis for Effective Teaching.* New York, NY: Routledge.

For more on attachments animals form with other animals:
Reaching the Animal Mind, Chapter 6.

For more about paying kids cash to get better grades:
Amanda Ripley. 2010. "Should Schools Bribe Kids?" *TIME,* April 10.

For a good demo of a puppy developing different relationships with two cats:
Orr, Joan. 2005. DVD. 48 mins. *Clicker Puppy.* Waltham, MA: Sunshine Books (available at www.clickertraining.com/store).

For tips on how to figure out whether you're reinforcing behavior you'd like to change:
Parsons, Emma. 2005. *Click to Calm, Healing the Aggressive Dog.* Waltham, MA: Sunshine Books, pp. 82–85.

Chapter 5: The Bad Stuff

For more on the unpredictable effects of punishing a behavior:
Sidman, Murray. 2000. *Coercion and Its Fallout.* Rev. ed. Boston, MA: Authors Cooperative. (available from The Cambridge Center for Behavioral Science, Beverly, MA. www.behavior.org/item.php?id=150)

For more on shaping with variable ratio schedules:
Broitman, Virginia. 2007. *The Shape of Bow Wow: Shaping Behaviors and Adding Cues.* DVD. 55 mins. Richmond, VA: North Star Canines & Co. (available at www.clickertraining.com/store)

For more on training dogs to love making noise:
Bertilsson, Eva, and Emelie Johnson Vegh. 2010. *Agility Right from the Start.* Waltham, MA: Sunshine Books, pp. 173–185.

To read the original paper that debunked dominance theory: Bradshaw, John W.S., Emily J. Blackwell, and Rachel A. Casey. "Dominance in Domestic Dogs—Useful Construct or Bad Habit?" *Journal of Veterinary Behavior: Clinical Applications and Research* (May/June 2009), 135-144. *Science Daily* presented a nice summary of the research: "Using 'Dominance' to Explain Dog Behavior Is Old Hat," *Science Daily,* May 25, 2009 at www.sciencedaily.com/releases/2009/05/090521112711.htm.

For some excellent examples of teaching distraction resistance with a moving dog:
Broitman, Virginia and Sheri Lippman. 2006. *Bow Wow Take II.* (combined on one DVD with *Take a Bow Wow!*). 71 mins. Richmond, VA, North Star Canines & Co. (available at www.clickertraining.com/store)

Chapter 6: A Confusion of Cues

For more on animals training us to respond:
Reaching the Animal Mind, Chapter 6.

For more on training llamas:
KPCT. "Polish, No Spit: Learning from Llamas." March 5, 2005. www.clickertraining.com/node/641.

For an excellent example of training a cue about intentions:
Parsons, Emma. 2005. *Click to Calm, Healing the Aggressive Dog.*
Waltham, MA: Sunshine Books, pp. 87–88.

To see how Nicole Murrey set up her experiment about poisoning a cue:
Murrey, Nicole. 2007. "The Effects of Combining Positive and Negative
Reinforcement during Training," Masters Thesis, University of North Texas.
http://reachingtheanimalmind.com/pdfs/ch_09/ch_09_pdf_05.pdf.

Websites:
For more on canine freestyle:

Attila Szkukalek as Charlie Chaplin routine:
https://www.youtube.com/watch?v=Vg3IrgfHXY0.

Michele Pouliot Seminars: www.cdf-freestyle.com.

Karen Pryor Academy's online course with Michele Pouliot on freestyle:
https://www.karenpryoracademy.com/canine-freestyle

Chapter 7: Expanding the Conversation

For publications about husbandry training at zoos, see the Animal
Behavior Management Alliance, which is the zoo keeper/trainer
association: www.theabma.org. They have a publication, called *Wellspring*.
Anyone interested in training can become an associate member and
receive the publication and visit the website.

For more on how even brief delays in marking behavior can negate
learning:
Grice, G.R., 1948. The relation of secondary reinforcement to delayed
reward in visual discrimination learning. *Journal of Experimental
Psychology*, 38, 271-282.

For Karen's history as a marine mammal scientist:
Pryor, K., 2014. "Historical Perspectives: A Dolphin Journey." *Aquatic
Mammals*, 40 (1), 104-115.

For the final product on the update of training for creativity, after many revisions:
Pryor, K., and Chase, S., 2014. "Training for Variable and Innovative Behavior." *International Journal of Comparative Psychology, Special Issue*, 27 (2), 361-368. (http://escholarship.org/uc/item/9cs2q3nr)

For Karen's most-cited scientific paper—on the dolphin who thought up her own tricks:
Pryor, K, Haag, R., and O'Reilly, J. 1969. "The Creative Porpoise: Training for Novel Behavior." *Journal of Experimental Analysis of Behavior*.

For Karen and Ken Ramirez's joint overview of modern animal training:
Pryor, K., and Ramirez, K., 2014. "Modern Animal Training: a transformative technology." In *The Wiley-Blackwell Handbook of Operant and Classical Conditioning*, First Edition, Frances K. McSweeney and Eric S. Murphy, Eds. John Wiley & Sons, Ltd.

For more on learning opportunities at three-day seminars with hands-on labs and an international staff of clicker experts: www.clickerexpo.com

Karen Pryor Academy offers intensive on-line courses, mentored by skilled faculty, on operant training and teaching for novice to professional levels across a variety of applications.
For information, course descriptions, schedules, or to take a free demo of any course: www.karenpryoracademy.com.

For shelter staff and volunteers:
www.karenpryoracademy.com/shelter-training-and-enrichment.
For professional-level education:
www.karenpryoracademy.com/dog-trainer-program.
For entry-level education:
www.karenpryoracademy.com/dog-trainer-foundations.
For puppy class planning for instructors:
www.karenpryoracademy.com/puppy-start-right.
For learning canine freestyle: www.karenpryoracademy.com/canine-freestyle.
For more on TAGteaching: TAGteaching.com

About Karen Pryor

From early childhood Karen Pryor was both a writer and a budding scientist. With a Cornell education and graduate studies in marine biology at the University of Hawaii and in zoology at New York University and Rutgers, she is the author of 20 articles and book chapters in the scientific literature, an edited volume on open-ocean dolphin studies, and 10 books for the general public. Her first of these, *Nursing your Baby*, sold over 2 million copies.

In 1963 she became head trainer at a new facility, Sea Life Park and the Oceanic Institute in Hawaii. Combining the work of Konrad Lorenz and the behavioral principles of B.F. Skinner, she pioneered in the development of modern, marker-based animal training.

After leaving Sea Life Park, Pryor published *Lads before the Wind*—an account of her early days as a dolphin trainer. She wasn't done with dolphins, though. Moving to New York City, she became scientific advisor to the tuna fishing industry. She also served as principal investigator on a National Marine Fisheries Service study of the behavior of dolphins in tuna nets. This research led to a joint project with marine mammalogist Kenneth S. Norris, PhD, contributing to and co-editing a textbook on studies of dolphin social organization. In 1984, President Reagan appointed her to a three-year term as Federal Commissioner on the Marine Mammal Commission, overseeing research and management of all marine mammals in US waters.

In 1984, Pryor published *Don't Shoot the Dog!* explaining the uses of positive reinforcement and marker-based training for human beings as well as animals. Now translated into 17 languages, the book has become a standard college text on reinforcement and a popular text for trainers of dogs and of zoo animals. *Reaching the Animal Mind,* published in 2009, is a follow-up.

Public demand for additional teaching materials led to Pryor's formation in 1992 of a publishing and video production company, Sunshine Books (later renamed Karen Pryor Clicker Training, or KPCT). In 2014, Pryor stepped out of her management role in the company to focus on research and spend more time with her family.

Karen Pryor has two sons and a daughter and seven grandchildren. She lives in Watertown, Massachusetts.